Social, Economic and Symbolic Perspectives at the Dawn of Metal Production

Edited by

Cecilia Conati Barbaro
Cristina Lemorini

BAR International Series 2372
2012

Published in 2016 by
BAR Publishing, Oxford

BAR International Series 2372

Social, Economic and Symbolic Perspectives at the Dawn of Metal Production

ISBN 978 1 4073 0962 0

BAR Publishing is the trading name of British Archaeological Reports (Oxford) Ltd.
British Archaeological Reports was first incorporated in 1974 to publish the BAR
Series, International and British. In 1992 Hadrian Books Ltd became part of the BAR
group. This volume was originally published by Archaeopress in conjunction with
British Archaeological Reports (Oxford) Ltd / Hadrian Books Ltd, the Series principal
publisher, in 2012. This present volume is published by BAR Publishing, 2016.

Printed in England

BAR
PUBLISHING

BAR titles are available from:

BAR Publishing
122 Banbury Rd, Oxford, OX2 7BP, UK
EMAIL   info@barpublishing.com
PHONE   +44 (0)1865 310431
FAX   +44 (0)1865 316916
www.barpublishing.com

TABLE OF CONTENTS

INTRODUCTION

Cecilia Conati Barbaro and Cristina Lemorini

The beginning of metal use in prehistoric communities has brought about important economic and social changes whose dynamics still remain partially unknown.

The transformations related to the "discovery" of metal artifacts have an early beginning: as a matter of fact, the first proof of copper treatment can be traced back to Neolithic in Europe and the Near East.

Since that time, such material has been gradually playing an important role in the life of farmers and herders. The use of metal tools has changed their habits and, slowly, the role that stone and osseous industries have played until that time.

The scarcity of metal finds in Copper Age sites, except some funerary contexts which reflect a specific social role, makes alternative sources essential in order to shed light on some important questions:

- Have metal artifacts been conceived as daily-life tools since the beginning or did they play an exclusively ritual and symbolic role, as status symbols?
- How long did stone and bone tools maintain their value of daily life tools?
- How far the way of making stone and osseous tools was affected by the use of metal implements?
- Did the introduction of metal artifacts modify the social meaning of some categories of stone/osseous tools? Has the introduction of metal artifacts restricted their social role?
- Did the introduction of morphologically "new" metal tools produce a change of the "making" only (gestures related to the use of different tools) or also of the "thinking" (cognitive categories associated to specific cultural choices)?

All these questions arise when we started to work on some Italian Copper age context (e.g. Conati Barbaro and Lemorini 2000, 2002; Conati Barbaro *et al*. 2002, 2010; Cristiani and Alaique 2005; Lemorini 2002, 2006; Lemorini and Massussi 2003) and we realize that the possible answers could be obtained only with a broader perspective. Therefore we wanted to open this debate to all scholars which are involved in research about Copper Age contexts in- and outside Europe.

The following papers represent a sort of preliminary answers to our main questions, reporting different case-studies and perspectives. Despite the variety of cases, all of them outline the importance of an integrated analysis of artifacts, considering pottery, metal, stone and osseous productions as inseparable aspects of economic and social choices.

This methodological setting clearly appears in the analysis of the Sardinian evidences offered by Melis *et al*. in the paper "The beginning of metallurgic production and the socioeconomic transformations of the Sardinian Eneolithic". The major changes which affected the Copper Age societies of the island, and which are clearly visible in funerary context, are deeply rooted in the gradual transformation of the Late Neolithic communities. If this could be related to the introduction of metal or not is a matter of discussion throughout the paper, analyzing the pottery, the osseous and lithic productions of some Sardinian sites. A general trend toward a "decline" of the Late Neolithic handcraft productions is observed in the island.

The paper of Conati Barbaro "First metals, last flints? The lithic productions of Central Italy at the end of the Neolithic. Some considerations" deals with the organization of lithic production at a time when copper was already known and used in everyday life, as attested by a growing evidence of metal artifacts, slags and crucibles in Neolithic settlements. Metal tools seem not to have replaced stone or osseous artifacts but they were probably improving the traditional productions, making easier and faster the manufacturing and maintenance of tool-kit. Nevertheless, technical changes are strictly related to social transformation, which could be related, besides the "metal innovation", also to new economic setting due to an intensification of agriculture and/or pastoral activities.

The intensification of production is clearly detectable in mining activities, as attested at the Gargano flint mines. As reported by Tarantini in his paper "A view from the mines. Flint exploitation in the Gargano (South-eastern Italy) and socio-economic aspects of raw materials procurement at the dawn of metal production.", flint mines are part of an integrated system; therefore any changes in extracting activity should be related to the transformations which affected the societies at the end of the Neolithic to the Copper Age. The idea of the economy as an integrated system has been clearly described by Karl Marx, as quoted by Tarantini, who wrote "production, distribution, exchange and consumption [...] are elements of a totality, differences within a unity: [...] *there is an interaction between the different components*" (Marx 1857-58).

In his paper Tarantini suggests that the flourishing of flint mines during Late Neolithic and Chalcolithic in Italian peninsula might be related to the raw material exploitation for bifacial tools such arrowheads which constitute, during Copper Age, the prevailing part of the stone kit of the burials. These types of lithic artefacts, together with daggers and long blade knifes, belong to a specialized craftsmanship, which contrasts with the expedient chipped artefacts that are especially present in the living contexts.

As Terradas *et al.* show in their paper "Producing for the Dead, Using while Alive: lithic tools production and consumption in the Late Neolithic of north-eastern Iberia.", the dichotomy between the lithic tools from burials and settlements in terms of technological "*savoir faire*" is well detectable in North-Eastern Iberia from the Late Neolithic (see also, for Central Italy, Conati Barbaro and Lemorini 2000, 2002; Conati Barbaro *et al.* 2002, 2010). This dichotomy is certainly related to the introduction of metal in the daily life of the Late-Neolithic and Chalcolithic communities; nevertheless, the social role of the two groups of artefacts - metal and lithic– is far from being clear. For example, long blade knifes, as stressed by Terradas *et al.* and Lemorini in her paper "Buried without metal: the role of lithic kit in Chalcolithic funerary contexts of the Marche region (Central Italy)", were "special" daily life tools, used for a very long time, probably throughout the life of their owners with which these items were buried. Instead, lithic daggers (see Lemorini paper) seem to be unused artefacts that played a symbolic role of prestigious objects related to socially prominent male individuals. Their amazing similarity to contemporaneous metal daggers found in similar burial contexts highlights that the "new" symbolic value of some special metal objects may have been transferred on lithic "analogues" in those communities in which this raw material had a "very rooted and not easily minable" social role.

Last but not least, the paper of Salazar *et al.* opens a window on the exploitation of metal in the "New World". Salazar *et al.* provide an interesting picture of the exploitation of copper mines of the highlands of Northern Chile by pre-Colombian civilizations. Combining data from technological analysis of heavy-duty tool-kit used for copper extraction with the quantitative and qualitative analysis of the space-use nearby the mines, the authors interpreted the increasing of mines exploitation in terms of changes in the socio-economic structure of the local communities whose mining traditional *expertise* was gradually assimilated in the Incan empire system of work organization and exploitation of the resources.

REFERENCES

Conati Barbaro, C. and C. Lemorini 2000. Oltre la tipologia: proposta per una lettura tecnologica e funzionale delle industrie litiche della prima età dei metalli. In M. Silvestrini (ed.), *Recenti acquisizioni, problemi e prospettive della ricerca sull'Eneolitico dell'Italia centrale*, 309-317, Arcevia.

Conati Barbaro, C. and C. Lemorini 2002. Riflessioni da una lettura integrata dei dati. In: A. Manfredini (ed.), *Le dune, il lago, il mare: una comunità di villaggio dell'età del rame a Maccarese,* Origines, 202-203, Firenze.

Conati Barbaro, C., Lemorini, C. and A. Ciarico 2002. Osservazioni sul potenziale interpretativo delle industrie litiche: un'applicazione a contesti del neolitico tardo in Italia centrale. In A. Ferrari and P. Visentini (eds.), *Il declino del mondo neolitico*, Quaderni del Museo Archeologico del Friuli Occidentale, 4, 167-176.

Conati Barbaro, C., Lemorini C. and E. Cristiani 2010. The lithic perspective: reading Copper age societies by means of techno-functional approach. *Human Evolution*, 25,1-2, 143-154.

Cristiani, E. and F. Alaique 2005. Selce o metallo? Approccio sperimentale all'analisi delle modalità di manifattura degli strumenti in materia dura animale presso Conelle di Arcevia (Ancona). Atti XXXVIII Riunione Scientifica I.I.P.P., 939-943.

Lemorini, C., 2002. Interpretazione funzionale dell'industria litica. In A. Manfredini (ed.). *Le dune, il lago, il mare: una comunità di villaggio dell'età del rame a Maccarese,* 188-199. Origines, Firenze, Istituto Italiano di Preistoria e Protostoria.

Lemorini, C. and M. Massussi 2003. Lo studio dei foliati in selce di Conelle di Arcevia: approccio tecno-funzionale, sperimentale e delle tracce d'uso. In A. Cazzella, M. Moscoloni and G. Recchia (eds.). *Conelle di Arcevia II. I manufatti in pietra scheggiata e levigata, in materia dura di origine animale, in ceramica non vascolari, il concotto*, 309-354. Università "La Sapienza", Roma.

Lemorini, C. 2006. Studio funzionale delle cuspidi di freccia delle tombe 1-5 del sito di Rinaldone (Viterbo). In A. Dolfini (ed.). La necropoli di Rinaldone (Montefiascone, Viterbo): rituale funerario e dinamiche sociali di una comunità eneolitica in Italia centrale. *Bullettino di Paletnologia Italiana* 95, 265-272.

Marx, K. 1857-58. *Grundrisse der Kritik der politischen Okonomie*. Rohentwurf, Berlin 1953.

FIRST METALS, LAST FLINTS? THE LITHIC PRODUCTIONS OF CENTRAL ITALY AT THE END OF THE NEOLITHIC. SOME CONSIDERATIONS

Cecilia Conati Barbaro

Dipartimento di Scienze dell'Antichità
Sapienza Università di Roma, Rome (Italy)
E-mail: cecilia.conati@uniroma1.it

In this paper I will focus on one of the many questions posed in the call for this book: if, and how far, the beginning of metal use affected lithic production. The first evidence of metal working and its use in Italy dates to the second half of the Vth mill. BC cal. Metal is mainly recorded in open air settlements (Figure 1), but it is also present in some cave sites. At present, we do not have evidence of copper objects as burial goods and this seems quite odd, as Copper Age metal artefacts come mostly from graves. In fact, the first structured cemeteries appear during the late Neolithic, when burial practices and tomb construction and disposal show an increasing evidence of social differentiation, as attested, for example, by the VBQ (Square Mouthed Pottery) cemeteries of Northern Italy, by the hypogean structures of Apulia or by the monumental stone slab tombs of Madonna delle Grazie at Rutigliano, still in Apulia. This picture changes during the Copper Age, when metal occurs rarely in settlement contexts if compared to the richer evidence from burial sites.

The emergence of copper represents a disruptive technological innovation, especially for its effects on the improvement of manufacturing other materials such as stone, bone and antler, rather than for the copper metalworking itself.

As a matter of fact, this new material – copper – should not be seen as an alternative to stone or bone/antler: as already pointed out by Rosen (1996), the replacement of the stone tool-kit with the metal one does not follow a simple unilinear process. Both technical systems are linked to different needs, procedures and goals, which are regulated by social and economic factors. The idea of "replacement" issues from the misconception of considering metallurgy as having the same value for the late Neolithic-early Copper Age communities and for the Bronze Age and later societies.

Without entering the discussion about the ways copper technology has been introduced in Italy – as it is well known, there is no general consent about this point: the three hypotheses consider either the French-chassean area, or the Transalpine regions or the Balkan area across the Adriatic sea as the cores of metal provenance for Italian peninsula – it should be pointed out that this process took place in a social context with enhanced social and cultural connections between communities, testified by strong affinities in material culture and increasing exchange networks. Obsidian from Lipari, Palmarola, Sardinia spreads, in different percentages, all over Italy, as well as good quality flint, e.g. Bedulian in the North-West, Lessinian in the North-East, Gargano in the Central-South. Is it therefore possible to perceive a change in the organization of the lithic production of the Late Neolithic communities connected to the introduction of copper?

As a case study I have chosen the Late-Final Neolithic sites of Central Italy for two main reasons: firstly because some of the first metal evidence comes from a few of these sites, secondly because the Tyrrenian side of Central Italy is rich in mineral sources, which have been exploited until modern times. Moreover, the first, well documented evidences of Copper Age communities of peninsular Italy are located along the Adriatic side of this region and these may be evaluated as terms of comparison in order to grasp changes in the long run.

The sites were selected according to the presence of quantitative analyses of lithic industries, which are not always as extensively published as pottery quantitative analyses are. Moreover, often only tools are described, while no account is taken for cores and débitage.

All sites are open-air settlements; burial sites were intentionally excluded to avoid any possible ritual/funerary implication in the composition of lithic assemblages.

Six sites are located in Tuscany (Podere Casanuova, Neto di Bolasse, Neto Via Verga 7, Neto Via Verga 5, La Consuma 1, Chiarentana), one in Umbria (Norcia), five sites in Marche (Cava Giacometti one, Saline di Senigallia, S. Maria in Selva, Calcinaia Serra S. Abbondio, Pianacci dei Fossi di Genga), four in Latium (Quadrato di Torre Spaccata, Valle Ottara, Casale di Valleranello, Casali di

Figure 1: Late-final neolithic sites: black dots: first metals evidences; white dots: sites quoted in the text.

Porta Medaglia), two in Abruzzi (Fossacesia, Settefonti). Almost all of them are radiocarbon dated and span from the second half of the Vth millennium BC to the first half of the IVth millennium BC cal. (Figures 2, 3).

Despite Central Italy being a largely mountainous land, with peaks rising above 2500 meters, the material culture of both Adriatic and Tyrrhenian sides during the Neolithic and even

more in the Copper Age, shows strong affinities that may prove long lasting connections between groups.

The wide valleys of Arno, Tiber and Aniene rivers have always been the preferred travel routes; the Nera river is a major route that connects Marche, via Umbria, to Latium. The natural ways often overlap with the ancient transhumance routes and with the present road system.

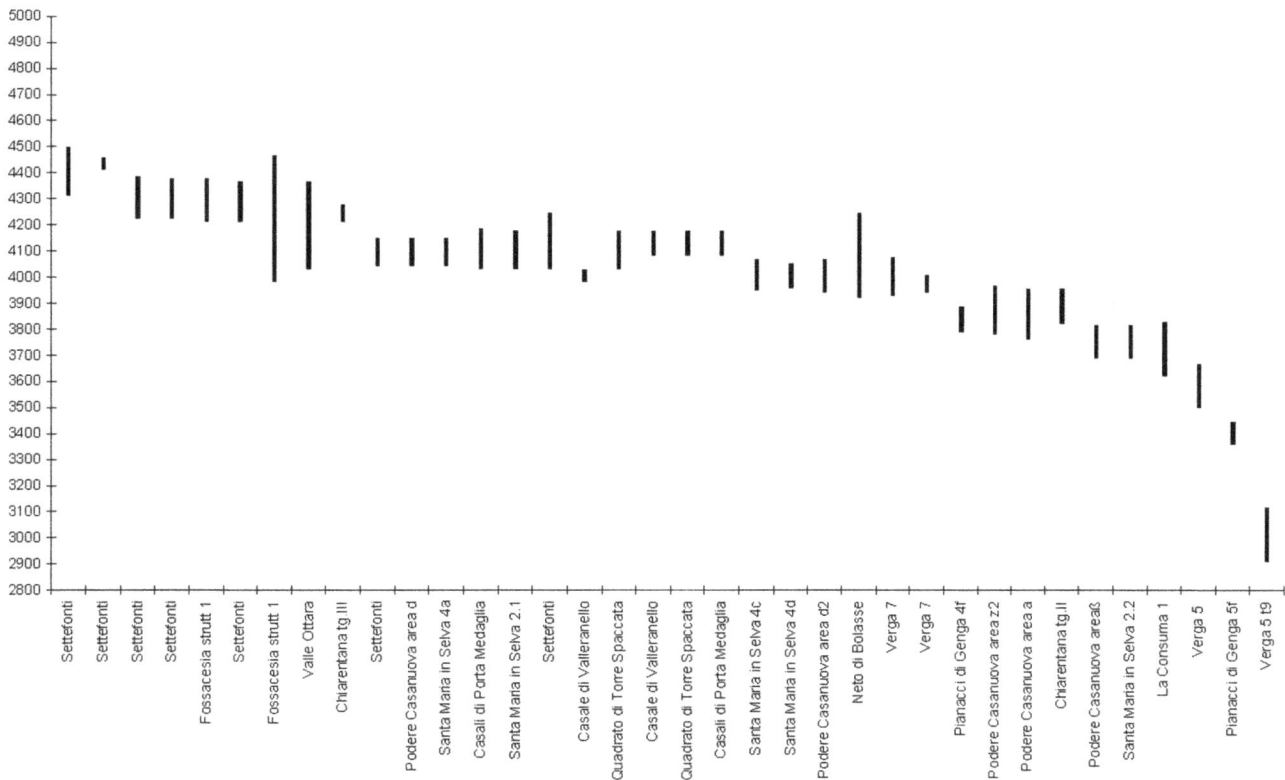

Figure 2: Absolute chronology of the Late-final Neolithic sites of central Italy quoted in this paper (cal.1 σ).

Therefore as observed by Puglisi, "the Apennines tie rather than divide" (Puglisi 1959).

What's new?

I will focus my analysis on three main aspects which represent different steps of the *chaîne operatoire*: raw material procurement; first working phases and cores; blanks selection and tools. We assume that all these steps involve different choices, depending on the organization of production, which reflect the socio-economic strategies of a community. Different raw materials at a site may indicate different energy investment for specialized or non specialized productions, way of access to the sources, circulation and exchange of exotic material. The specialization level of the tool-kit and the tools function, if trace wear analyses are available, may be indicative of some change in the economic basis of a human group.

Raw materials
Raw materials are usually collected on a local basis (Figure 4). Flint is the most used raw material, followed by jasper in some Tuscanian sites. Good quality, fine grained flint is mostly chosen to obtain blades and/or bladelets, but lower quality flint types are also used mainly for flakes and occasional tool types.

Non local, good quality flint is attested in a few sites as Verga 7, Verga 5, Fossacesia, Casali di Porta Medaglia: in these two last sites a Gargano source can be assumed.

Flint characterization is another topical question, which has been rather systematically treated only in recent times. As an example, research on the Florentine area led to the identification of different procurement zones. At Neto Via Verga 7 and Verga 5 local flint from the alluvial deposits is mostly attested (70%), while a small amount of exogenous flint comes from the Umbro-Marchigiano Domain, 100-150 km away (Martini *et al.* 2006). The lithic raw material of Settefonti (Abruzzo) has been obtained from outcroppings in the Gran Sasso area, not far from the site (Danese 2003).

Although in variable quantity, obsidian is attested too: Chiarentana (51,5%) and Quadrato di Torre Spaccata (36,55%) have a considerable amount of obsidian, which is rather unusual if compared to the other sites. Characterization analysis has been made only on some samples: Lipari is by far the most represented source, often associated with Palmarola, which is the second source in order of importance. Obsidian from the Sardinian source of Mount Arci occurs only in few Tuscanian sites. Specimens coming from all of the three sources are only attested at Quadrato di Torre Spaccata.

Steatite is another external lithic resource which is well attested at La Consuma and in the Roman area at Quadrato di Torre Spaccata and Casali di Porta Medaglia. Primary sources of this soft stone are located in Tuscany, in the Livorno district and in Garfagnana, a mountain area in the Northern part of the region.

The data reported above show a remarkable intensification

Sites	BP	bC cal 1σ	Lab
Settefonti	5535±90	4490-4320	R-774
Settefonti	5480±65	4450-4420	R-772
Settefonti	5470±75	4380-4230	R-401
Settefonti	5460±70	4370-4230	R-771
Fossacesia strutt 1	5430±120	4370-4220	R-878
Settefonti	5425±70	4360-4220	R-770
Fossacesia strutt 1	5420±210	4460-3990	F-30
Valle Ottara	5398±145	4360-4040	Pi-28
Chiarentana tg.III	5370±50	4270-4220	Beta-153649
Settefonti	5350±75	4140-4050	R-402
Podere Casanuova area d	5350±70	4140-4050	Bln- ?
Santa Maria in Selva 4a	5322±45	4140-4050	LTL 1486A
Casali di Porta Medaglia	5300±50	4180-4040	OxA-7707
Santa Maria in Selva 2.1	5300±40	4170-4040	LTL-1484A
Settefonti	5295±70	4240-4040	R-769
Casale di Valleranello	5280±65	4020-3990	OxA-4328
Quadrato di Torre Spaccata	5280±50	4170-4040	OxA-6050
Casale di Valleranello	5280±45	4170-4090	OxA-8916
Quadrato di Torre Spaccata	5270±50	4170-4090	OxA 6049
Casali di Porta Medaglia	5270±50	4170-4090	OxA-7706
Santa Maria in Selva 4c	5225±45	4060-3960	LTL-1487A
Santa Maria in Selva 4d	5202±45	4045-3965	LTL-1488A
Podere Casanuova area d2	5200±60	4060-3950	Paris Sud- ?
Neto di Bolasse	5200±150	4240-3930	UD-184
Verga 7	5190±70	4070-3940	?
Verga 7	5170±40	4000-3950	?
Pianacci di Genga 4f	5116±65	3880-3800	LTL-1490A
Podere Casanuova area z2	5080±70	3960-3790	Bln- ?
Podere Casanuova area a	5040±60	3950-3770	Bln- ?
Chiarentana tg.II	5040±40	3950-3830	Beta-145962
Podere Casanuova areaß	5000±70	3810-3700	Paris Sud- ?
Santa Maria in Selva 2.2	4995±55	3810-3700	LTL-1485A
La Consuma 1	4920±130	3820-3630	UTC-820
Verga 5	4790±80	3660-3510	?
Pianacci di Genga 5f	4723±50	3440-3370	LTL-1489A
Verga 5 t9	4427±65	3110-2920	LTL-1481A

Atmosferic data from Reimer et alii (2004); OxCal v3.10 Bronk Ramsey (2005); cub 2.5 sd:12 prob usp [chron]

Figure 3: Radiocarbon dates of the sites quoted in the text.

of raw material circulation and exchange. Different kinds of stones, often co-existing at sites, were sought at short, middle and long distance from settlements.

Does this evident increase of exchange activities respond to a major demand for high quality goods? Or is it related to forming and maintaining enhanced social relations?

First reduction phases and cores
Although the general picture outlined from the current data appears rather diversified, we may point out some main trends of the *chaine opératoire* (Figure 5):

- cores are often few, small or overexploited;
- multidirectional flake cores are the most attested items;
- bladelet cores are less documented; blade cores are rare;

Sites	flint	diasper	obsidian	limestone	quarz	quarzite
Podere Casanuova	12,4	81,3	5,97*	-	0,39	-
Neto di Bolasse	70,93	1,16	6,97**	20,93	-	-
La-Consuma 1	94	3	0,45***	-	-	2
Chiarentana	47,9	-	51,5***	-	0,6	-
Verga 7	90	x	x**	1	-	-
Verga 5	90	x	x**	1	-	-
Norcia	98	-	2	-	-	-
Cava Giacometti	100	-	-	-	-	-
Saline di Senigallia	x	-	x	-	-	-
Santa Maria in Selva area 3	x	-	2,6	x	-	-
Santa Maria in Selva area 2	x	-	x	x	-	-
Santa Maria in Selva area 1	x	-	-	-	-	-
Calcinaia S.Abbondio	99,3	-	0,7	-	-	-
Pianacci di Genga	100	-	-	-	-	-
Quadrato di Torre Spaccata	61,76	0,17	36,55****	1,52	-	-
Valle Ottara	99	-	1	-	-	-
Casale di Valleranello	100	-	-	-	-	-
Casali di Porta Medaglia	x	-	-	-	-	-
Fossacesia	-	-	9,89***	-	-	-
Fossacesia features. 2-9	94,3	-	5,7	-	-	-
Settefonti	92	-	8***	-	-	-

Figure 4: Raw materials frequencies (values are in percentages). x = presence, if quantity is not quoted in publications. Obsidian sources: *Arci; ** Arci and Palmarola; *** Lipari and Palmarola, **** Lipari, Palmarola and M. Arci. Italic=sites with metals.

Sites	cores	%	débitage	%	tools	%
Podere Casanuova	11	4,4	178	71,8	59	23,8
Neto di Bolasse	3	3,5	66	76,7	17	19,8
La Consuma 1	10	0,6	1657	93,4	108	6,1
Chiarentana	7	4,2	118	71,5	40	24,2
Norcia	22	6,5	136	39,9	183	53,7
Cava Giacometti	73	1,6	3497	78,6	879	19,8
Saline di Senigallia	-	-	-	-	264	-
Santa Maria in Selva area 3	31	2,3	1043	77,4	274	20,3
Santa Maria in Selva area 2	100	-	?	-	2000	-
Santa Maria in Selva area 1	18	-	45	-	102	-
Calcinaia S Abbondio	17	-	399	-	159	-
Quadrato di Torre Spaccata	12	2	443	75	135	23
Fossacesia strutt 2-9	90	3,7	1606	66,2	730	30,1
Settefonti	24	2,9	499	60,2	217	26,2

Figure 5: Frequencies of main technological classes; italic = sites with metals.

- all the obsidian cores are for bladelets;
- the correspondence between raw material, core and blanks is hardly detectable: for example, it's hard to say whether external raw materials (apart from obsidian) are used to make blades or bladelets or specific tools;
- the first phases of core reduction are poorly attested, apart from a few cases of dedicated flaking areas (Cava Giacometti, Saline di Senigallia, S. Maria in Selva area 2);
- crested blades and rejuvenation elements, attested at some sites (e.g. Norcia, Cava Giacometti), may indicate an *in situ* blade production;
- in some sites there are no *débitage* elements made of external raw materials: for example at QTS non local good quality flint (maybe from the Gargano area) is attested only by finished products. Functional analysis showed that these elements were used for different activities and for a long time. In some cases we may admit different raw material economy (*sensu* Perlès 1991), with the introduction on site of non local finished blanks which were maintained over time.
- pressure flaking technique could be assumed at some sites (Calcinaia, Fossacesia, Settefonti) by the presence of flint and obsidian bladelets with regular margins and parallel edges.

Blanks and tools

A series of common traits can be outlined also for these categories:

- blades and bladelets seem to be the preferred products. Some of them have been retouched, but often they were used as natural blanks. Blade and bladelets are frequently broken;
- as far as the tool-kit composition is concerned (Figure 6), all sites except Podere Casanuova show a clear prevalence of Substratum *sensu* Lapace (mainly sidescrapers, notches and denticulates). Flakes are the most frequent blanks for this group of tools at many sites (Podere Casanuova, Neto di Bolasse, La Consuma 1, Chiarentana, Calcinaia), but in some of the other sites blades are more common (Norcia, Cava Giacometti, Fossacesia). The retouch is almost always marginal and partial, not very accurate, as for the choice of blanks: this behaviour may suggest the use of these tools for occasional activities.
- burins and end-scrapers are very rare or totally absent. This is a significant fact if we compare the typological structure of the previous Neolithic phases, when a wide typological variety of burins occurred.
- the group of Abrupt types (*sensu* Laplace) occurs quite frequently, but the percentage *ratio* among tool types differs from site to site: truncations and drills are more attested than backed elements while geometrics occur only at Verga 7, Fossacesia and Settefonti;
- foliated tools are the third group in order of importance. The shape of these tools is greatly variable from site to site: a strong link of these tools with cultural identities has been suggested by many authors. In a

recent paper Baglioni, Martini and Volante (Baglioni *et al.* 2005) consider chronological factors to explain the morphological and typological variations of these tool types which occurred during the Late and the Final Neolithic in Central Italy.

To sum up, blades and more often bladelets are the preferred blanks. Beside the numerical data, this is confirmed by use-wear analysis which has been carried out in some sites (Santa Maria in Selva, Neto Via Verga 7 and 5, Quadrato di Torre Spaccata). Blades and bladelets were used for many different functions: skin treatments, wood, antler and stone working, cutting of herbaceous plants, and, to a lesser extent, of cereals. Blades and bladelets show a prolonged use, mostly for those made of non local raw materials, like obsidian (Lemorini *et al.* 1995). At Quadrato di Torre Spaccata, for example, obsidian blades show a remarkable polish development attesting a heavy use for many activities, mainly skin tanning and a long functional life (Lemorini *et al.* 1995).

Traditional tool types as end-scrapers and burins are rare, while the most represented tools are truncations, geometrics and, above all, foliated.

Ad hoc tools, mainly on flakes, are also documented; functional analysis indicate occasional, non prolonged use of these tools (Conati Barbaro *et al.* 2002).

If we look at coeval context outside of the sample area chosen as case-study, a good term of comparison for techno-functional observations is Bannia (Friuli Venezia Giulia, Northeastern Italy), a settlement dated to the last phase of Square Mouthed Pottery (VBQ) culture (5615±45 BP, 4490-4360 a.C. cal.) where a pure copper awl has been recovered. The lithic industry is orientated to blades and bladelets production, mainly made of non local raw materials. Blades and bladelets are the preferential blanks for tools, especially for specialized types as end-scrapers, foliated and points. Flakes are used as unretouched blanks, taking advantage of natural margins for different unspecialized activities (Dal Santo 2005; Lemorini 2005).

As a general remark, we can outline a Late Neolithic trend to focus the tool production on few specialized types (foliates, geometrics), to use unspecific blanks, mainly on local raw materials, for occasional uses and to use blades and bladelets for a wide range of daily activities, maintainig their functional efficiency by resharpening. This trend becomes a common practice during the Copper Age.

As an example we can quote the site of Le Cerquete-Maccarese, where a techno-functional approach has been applied in the lithic analysis (Conati Barbaro 2002; Conati Barbaro and Lemorini 2000, 2002; Conati Barbaro *et al.* 2010). Two technological strategies related to different raw materials have been observed: non local flint has been brought to the site as an end-product, namely as blades, which were heavily and long used as multi-purpose tools.

Sites	burins	end-scrapers	abrupt-retouch tools	foliates	substratum	outil écaillées	campignan tools
Podere Casanuova	-	3,4	11,9	47,5	32,2	1,7	1,7
Neto di Bolasse	5,9	-	5,9	5,9	52,9	17,6	11,8
La Consuma 1	0,92	3,7	4,6	14,8	60,2	11,1	4,6
Chiarentana	2,5	2,5	15,0	-	22,0	10,0	-
Verga 7	x	x	xx	-	xxx	xx	-
Verga 5	x	x	xx	xx	xxx	x	-
Norcia	-	4,9	5,5	13,1	71,0	4,9	0,5
Cava Giacometti	6,14	3,5	16,3	1,7	41,9	-	2,0
Santa Maria in Selva area 3	5	x	xx	xx	xxx	x	x
Santa Maria in Selva area 2	x	x	xx	xx	70,0	x	x
Santa Maria in Selva area 1	x	x	xx	xx	xxx	x	-
Saline di Senigallia	-	x	>10	x	70,0	-	-
Calcinaia S. Abbondio	1,3	2,0	18,8	3,9	66,5	6,7	-
Pianacci di Genga	-	-	-	-	-	-	-
Quadrato di Torre Spaccata	3,7	6,7	17,0	10,4	55,6	0,7	-
Fossacesia 1	2	1,7	8,2	1,2	87,0	-	-
Fossacesia 10	4	4,0	18,1	1,0	73,0	-	-
Fossacesia strutt 2-9	1,6	3,7	9,7	2,1	81,5	0,3	0,1
Settefonti	1,4	1,9	22,8	13,1	57,3	3,7	-

Figure 6: Frequencies of tool types groups according to Laplace's typology. Numbers indicate percentages. Unquantified values: x = scarce, xx = medium, xxx = abundant.

Non specialized and *ad hoc* tools were mainly on flakes, generally produced with local flint. However, local flint is also used for specialized tools as bifacial tools, projectile points and geometrics.

Were these the symptoms of some changes in the chipped stone productions possibly connected to metal use?

The economic basis

How far the new asset of lithic industries during Final Neolithic is affected either by the economic basis or by cultural tradition or by social reasons?.

If we look at the faunal and botanical data, the framework appears quite dissimilar from site to site. This can be related to sample preservation or selection criteria, different ecological settings and, of course, to cultural reasons.

As an example, the Tuscan villages of Podere Casanuova and Neto di Bolasse, both located in humid environments, feature very different faunal assemblages. Pigs are the most represented species in the first site, while in the second one caprovines are more frequent. The botanical data indicate the practice of cereal and legumes cultivation as well as the use of wild plants at Podere Casanuova; pollen records at Neto di Bolasse show the presence of wooded areas along with clearings opened by man for grazing animals, while cultivated plants are absent.

Caprovines prevail at Cava Giacometti and Quadrato di Torre Spaccata, while at Norcia and Santa Maria in Selva domestic cattle is the most important species.

When reported by publications, there is a general trend in killing adult animals rather than young or juveniles, mainly for cattle and caprovines, which is a custom traditionally linked to the exploitation of secondary products of animal husbandry. The idea of a major change to pastoralism at the end of Neolithic is not supported by the archaeozoological evidences: this is a possibility for a few cases, but cannot be seen as a general trend. The increasing importance of livestock as multipurpose, long-lasting goods can indirectly be proved by tools used for milk processing, as fragments of strainers (Valle Ottara) and pierced lids (Norcia), or for spinning and weaving, as spindle whorls and loom weights (Podere Casanuova, Quadrato di Torre Spaccata), altough this equipment could also be used for vegetal fibres, both wild and domesticated (e.g. flax at Settefonti).

Hunting is variously attested at many sites, but it is a relevant activity only in a few of them (Cava Giacometti 31%; S. Maria in Selva 21,2%; Valle Ottara 67,5%, Norcia 16,4%) (Wilkens 1993), although this particular choice seems hardly referrable to the ecological conditions. As a matter of fact Valle Ottara, Norcia and Cava Giacometti are inland sites, in hilly, wooded landscapes, which were the best habitat for red and roe deer, but other sites located in similar habitats do not show similar percentage of wild fauna.

According to the archaeobotanical evidence, agriculture

7

was still an important economic activity; in addition to the previously quoted sites, Quadrato di Torre Spaccata and Settefonti attest the cultivation of cereals (*Triticum* and *Hordeum*) and legumes. Moreover, flax is documented at Settefonti. Different percentages of various species may vary from site to site also depending on environmental conditions. Altough the data are so far quite scanty, there is no clear change in the economic basis, which is very similar to the previous phases of the Neolithic. Archaeobotanical analysis (pollen and charcoal) shows that the Campagna Romana (the countryside around Rome) featured a landscape composed of wooded areas and meadows; moreover at Quadrato di Torre Spaccata and Casali di Porta Medaglia there is a high percentage of evergreen vegetation, the growth of which can be favoured by human actions as burning and grazing degradation (Celant 2000, 2002). This general framework slightly contrasts with functional observations carried out on lithic tools: in fact, harvesting is not well attested, as one could expect. Collecting plants other than cereals appears instead to be a more common activity (Conati Barbaro *et al.* 2002).

First metals

It is worth noting that metal objects or working refusals are present in sites where is attested either an intense flaking activity, as at Santa Maria in Selva, or a wide range of good-quality raw materials, as external flint and obsidian at Fossacesia.

According to Skeates, throughout the Neolithic some sites can be distinguished as exchange junctions because of the presence of different types of exotic raw materials (Skeates 1992); to these I would add others where an intensification of production is attested. The first evidence of metal working is only recorded in some of these sites.

As already pointed out by Pearce (2000, 2007), as soon as metal appears in Italy, awls, rods and points are the most frequent artefacts besides copper slags and fragments of crucibles. Having discussed a range of possible uses of these small objects (tattooing, basketmaking, leatherworking, flint pressure flaking), Pearce suggests that they could be interpreted as retouchers connected to the coeval onset of bifacial flintwork. This is an intriguing hypothesis that could be broadened by including the use of these objects as points of copper punch for blade pressure flaking. In his recent PhD thesis, Guilbeau (2010, unpubl.) has taken into account the Neolithic and Calcolithic big blades of Italy to analyze the technological traces of their production technique. As far as Central Italy is concerned, he observes that big blades appear only at the end of Neolithic and rarely during the Copper Age. He recognize the use of the lever at the site of Santa Maria in Selva. This is particularly remarkable because one of the first evidences of copper working and use comes from this site.

However, the first metal use in Central Italy, but more generally in the whole peninsula, is attested not only

by the products of the so-called *"trinket metallurgy"* (Ottaway 2001: 103), but also by evidences of some steps of the *chaîne opératoire* (crucibles, slags*)*. We may assume a craft specialization, but only in the final stages of the sequence; mineral ores could have been obtained from outside the region by way of exchange, notably in the areas far from mineral sources. On the other hand, the production stages of smelting, refining, alloying and casting require the knowledge and skills of a specialist. Nevertheless, the dearth of metal working at these sites could also be related to conservation factors: as noted by Ottaway and Roberts (2008) "the archaeological and archaeometallurgical evidence for early smelting can be ephemeral. The careful processing and smelting of highly concentrated copper carbonate ore could have meant that there was little gangue attached to the mineral particles and thus only minimal amounts of slag would have formed. This could explain the dramatic under-representation of copper slags in certain regions and periods…".

The intensification of some specific productions (long blades, bifaces) and activities (prospection and extraction of good-quality flint and other raw materials such as obsidian) clearly points out a growing demand and circulation of high quality goods.

The introduction of a new material – copper – may be connected to different meanings and roles:

- *symbolic*: colour and brightness are sense properties which are value-added characteristics of copper; as attested by ethnographic and historical data from Africa and America, metal working is highly permeated with cosmological and mythological symbolism, often related to gender, reproduction and life cycle spheres (see, as an example, Falchetti 2003; Haaland 2004; Reid and MacLean 1995). As Robb points out (2007) aesthetic/symbolic changes took place during the Late Neolithic and the following Copper Age also affecting pottery production: metallurgy could have played a substantial role in developing this new attitude, to enter what Robb calls "the Age of Shine" (Robb 2007: 320).
- *functional*: copper tools, even if only awls or points, may simplify other raw material processing, as flint, bone, antler. This improvement is clearly documented since the beginning of the Copper Age: as a matter of fact, functional analysis of artefacts made out of hard animal materials from Copper Age sites of Central Italy (Conelle di Arcevia, Camerano, Fontenoce, Maddalena di Muccia) has been made and re-sharpened by means of metal tools, as shown by microtraces preserved on their surfaces (Cristiani and Fecchi 2003; Cristiani and Alhaique 2005). The use of metal should have improved and fastened the processing of hard animal tissues.

Metal tools were used for butchering animals and taking the flesh off as indicated, for instance, by the traces on the bones of a horse at Maccarese village (Latium). The horse was buried, together with two dog pups, in a pit just outside the centre of the settlement (Manfredini 2002; Curci and Tagliacozzo 1994, 2002). Functional and symbolic values are here closely related, given the clear ritual meaning of this burial feature.

Going back to the main question if and how lithic production has been affected by metal use, it can be observed that the above mentioned evidence from late Neolithic sites of Central Italy attest some general trends in exploiting and using lithic raw materials that could be connected to meeting new social demands (economic activities, social relations, etc.), and to an increased mobility. A progressive impoverishment of the typological set can be detected from the end of the Neolithic onwards, while a specialization on few types becomes more evident. Among these types arrow heads prevail, suggesting a shifting of some lithic categories into a more symbolic sphere connected to the individual identity. However, these major changes cannot be explained as a direct consequence of metal introduction, but rather as adjustments to a substantial social change, which involve the individual, the familiar, the communal, the regional and interregional spheres. As previously mentioned, at this chronological level, lithic and metal productions should not be seen as alternative or opposed, but as complementary chaînes opératoires.

References

Baglioni, L., Martini, F. and N. Volante 2005. Le industrie litiche del Neolitico recente e finale delle Marche. Atti XXXVIII Riunione Scientifica IIPP, 279-293.

Baglioni, L., Martini, F. and N. Volante 2008. Identità, variabilità e interazioni nei complessi litici tra V e III millennio a.C.: evoluzione e tendenze in industrie della Toscana e delle Marche. *Bullettino di Paletnologia Italiana* 97, 91-126.

Celant, A. 2000. Nuovi dati archeobotanici su ambiente e agricoltura nel Neolitico del Lazio: un esempio dalla Campagna Romana. In A. Pessina and G. Muscio (eds.), *La neolitizzazione tra oriente e Occidente*, 355-363. Udine.

Conati Barbaro, C. 2002. L'industria litica. Analisi tipologica e tecnologica. In A. Manfredini (ed.), *Le dune, il lago, il mare: una comunità di villaggio dell'età del rame a Maccarese*, Origines, 167-187. Firenze.

Conati Barbaro, C. and C. Lemorini 2000. Oltre la tipologia: proposta per una lettura tecnologica e funzionale delle industrie litiche della prima età dei metalli. In M. Silvestrini (ed.), *Recenti acquisizioni, problemi e prospettive della ricerca sull'Eneolitico dell'Italia centrale*, 309-317. Arcevia.

Conati Barbaro, C. and C. Lemorini 2002. Riflessioni da una lettura integrata dei dati. In: A. Manfredini (ed.), *Le dune, il lago, il mare: una comunità di villaggio dell'età del rame a Maccarese*, Origines, 202-203. Firenze.

Conati Barbaro, C., Lemorini, C. and A. Ciarico 2002. Osservazioni sul potenziale interpretativo delle industrie litiche: un'applicazione a contesti del neolitico tardo in Italia centrale,). In A. Ferrari and P. Visentini (eds.),

Il declino del mondo neolitico, Quaderni del Museo Archeologico del Friuli Occidentale, 4, 167-176.

Conati Barbaro, C., Lemorini, C. and E. Cristiani 2010. The lithic perspective: reading Copper age societies by means of techno-functional approach. *Human Evolution*, 25,1-2, 143-154.

Cristiani, E. and F. Fecchi 2003. I manufatti in materia dura animale: l'inquadramento tipologico ed i risultati dell'analisi tecno-funzionale. In A. Cazzella, M. Moscoloni and G. Recchia (eds.), *Conelle di Arcevia. Tecnologia e contatti culturali nel Mediterraneo Centrale fra IV e III Millennio A. C. II. I Manufatti in pietra scheggiata e levigata, in materia dura animale, in ceramica non vascolare; il concotto*, 423-502. Edizioni Stampa dell'Ateneo, Roma.

Cristiani, E. and F. Alhaique 2005. Selce o metallo? Approccio sperimentale all'analisi delle modalità di manifattura degli strumenti in materia dura animale presso Conelle di Arcevia (Ancona). Atti XXXVIII Riunione Scientifica IIPP, 939-943.

Curci, A. and A. Tagliacozzo 1994. Il pozzetto rituale con scheletro di cavallo dall'abitato eneolitico di Le Cerquete-Fianello (Maccarese, RM). Alcune considerazioni sulla domesticazione del cavallo e la sua introduzione in Italia. *Origini* XVIII, 297-350.

Curci, A. and A. Tagliacozzo 2002. Il pozzetto rituale con scheletro di cavallo dall'abitato eneolitico di Le Cerquete-Fianello (Maccarese, Fiumicino. In A. Manfredini (ed.), *Le dune, il lago, il mare: una comunità di villaggio dell'età del rame a Maccarese,* Origines, 238-245. Firenze.

Danese, F. 2003. Approvvigionamento della selce nei settori marsicano e aquilano dell'Abruzzo. Dati preliminari. Atti XXXVI Riunione Scientifica IIPP, 605-609.

De Grossi Mazzorin, J. and C. Minniti 1995. I resti faunistici dell'insediamento di Quadrato di Torrre Spaccata nel contesto delle economie di allevamento del Neolitico finale ed Eneolitico in Italia centrale. *Origini*, XIX, 287-295.

Eliade, M. 1956. *Forgerons et Alchimistes*. Flammarion, Paris.

Falchetti, A.M. 2003. The seed of life: the symbolic power of gold-copper alloys and metallurgical transformations. In J. Quilten, and J.W, Hoopers (eds.), *Gold and Power in Ancient Costa Rica, Panama and Colombia*, Dunbarton Oaks, Washington DC, 345-381.

Guilbeau, D. (unpubl.) - Les grandes lames et les lames par pression au levier du Néolithique et de l'Énéolithique en Italie. Thèse de Doctorat. Université Paris Ouest, sous la direction de C. Perlès.

Haaland, G. R. 2004. Smelting iron: Caste and its symbolism in south-western Ethiopia'. In T. Insoll (ed.), *Belief in the Past.*, British Archaeological Reports, International Series, 1212, 75-86, BAR Publishing, Oxford.

Manfredini, A. (ed.) 2002. *Le dune, il lago, il mare: una comunità di villaggio dell'età del rame a Maccarese.* Origines, Firenze.

Manfredini, A, Fugazzola, M.A, Sarti, L, Silvestrini, M, Martini, F, Conati Barbaro, C., Muntoni, I, Pizziolo, G.

and N. Volante 2009. Adriatico e Tirreno a confronto: analisi dell'occupazione territoriale tra il Neolitico finale e l' Età del Rame in alcune aree campione dell'Italia centrale. *Rivista di Scienze Preistoriche*, LIX, 115-180.

Marini, F., Ghinassi, M. and B. Moranduzzo 2006. Caratterizzazione degli areali e modalità di raccolta della materia prima litica in area fiorentina dal Paleolitico all'età del Bronzo. Atti XXXIX Riunione Scientifica IIPP, 299-313.

Ottaway, B. 1982. *Earliest copper artifacts of the Northalpine region: their analysis and evaluation.* Schriften des Seminärs für Urgeschichte der Universität Bern, Heft 7, Bern.

Ottaway, B. 2001. Innovation, production and specialization in early prehistoric copper metallurgy. *European Journal of Archaeology*, 4(1), 87-112

Pearce, M. 2000. What this awl means: understanding the earliest Italian metalwork. In Ridgway D., Serra Ridgway, F., Pearce, M., Herring, E., Whitehouse, R., Wilkins, J. (eds.) *Ancient Italy in its Mediterranean Setting: Studies in Honour of Ellen Macnamara.* Accordia, 4, London, 67-73.

Pearce, M. 2007. *Bright Blades and Red Metal. essays on North Italian prehistoric metalwork.* Accordia, 14, London.

Perlès, C. 1991. Economie des matières premières et économie du débitage: deux conceptions opposées? In *25 ans d'études technologiques en préhistoire : bilan et perspectives.* Rencontres internationales d'archéologie et d'histoire d'Antibes, 11, 35-45.

Puglisi, S. 1959. *La civiltà appenninica.* Sansoni: Firenze.

Reid, A. and R. MacLean 1995. Symbolism and the Social Context of Iron Production in Karagwe. *World Archaeology,* 27, 144-161.

Robb, J. 2007. *The Early Mediterranean Village.* Cambridge University Press, Cambridge.

Roberts, B. 2008. Migration, Craft Expertise and Metallurgy: Analysing the 'Spread' of Metal in Western Europe. *Archaeological Review from Cambridge*, 23, 2, 27-45.

Rosen, S.A. 1996. The Decline and Fall of Flint. In G.H. Odell (eds.), *Stone Tools: Theoretical Insights into Human Prehistory*, Plenum Press, New York, 129-155.

Skeates, R. 1993. Early metal-use in the central Mediterranean region. *Accordia*, 4, 5-48.

Visentini, P. ed. 2005. *Bannia-Palazzine di sopra. Una comunità preistorica del V Millennio a.C..* Quaderni del Museo archeologico del Friuli Occidentale, 5, Pordenone.

Wilkens, B. 1987. La fauna del villaggio neolitico di S.Maria in Selva. *Picus*,VII, 169-194.

Wilkens, B. 1991 Il ruolo della pastorizia nelle comunità preistoriche dell'Italia centro-meridionale. *Rivista di Studi Liguri*, 57, 81-94.

Wilkens, B. 1993. Etat des données archéozoologiques sur la chasse en Italie centrale et méridionale du Nèolithique à L'Age du bronze. Exploitation des animaux sauvages à travers le temps, XIII Rencontres Internationales

d'Archéologie et d'Histoire d'Antibes, APDCA, 261-274.Juan-les-Pins.

Tuscany

Podere Casanuova
Aranguren, B.M., Ducci, S. and P. Perazzi 1991. Il villaggio neolitico di Podere Casanuova (Pontedera, Pisa). *Rivista di Scienze Preistoriche*, XLIII, 155-239.

Neto di Bolasse
Sarti, L. 1985. L'insediamento neolitico di Neto di Bolasse (Sesto Fiorentino, Firenze). *Rassegna di Archeologia*, 5, 63-117.

La Consuma 1
Castelletti, L., Martinelli, M.C., Maspero, A. and A. Moroni 1992. Il sito neolitico della Consuma 1 (Pieve S. Stefano, Arezzo). *Rivista di Scienze Preistoriche*, XLIV, 43-114.

Chiarentana
Cuda, M.T. 2001. Il sito di Chiarentana (Chianciano Terme) e le ultime manifestazioni neolitiche nella Toscana meridionale interna. Atti XXXIV Riunione Scientifica IIPP, 367-382.

Cuda, M.T. 2002. Il Neolitico recente di Chiarentana (Chianciano Terme). In A. Ferrari and P. Visentini (eds.), *Il declino del mondo neolitico.* Quaderni del Museo Archeologico del Friuli Occidentale, 4, 447-452.

Verga 5, Verga 7
Volante, N. 2003. Neto-Via Verga (Sesto Fiorentino). La produzione vascolare dell'area I. *Rivista di Scienze Preistoriche*, LIII, 375-504.

Baglioni, L., Martini, F. and N. Volante 2008. Identità, variabilità e interazioni nei complessi litici tra V e III millennio a.C.: evoluzione e tendenze in industrie della Toscana e delle Marche. *Bullettino di Paletnologia Italiana*, 97, 91-126.

Umbria

Norcia
Corridi, C. and A. Moroni 1993. I materiali della capanna di Norcia conservati al Museo Archeologico di Perugia: industria litica, ossea e resti faunistici. *Bullettino di Paletnologia Italiana*, 84, 381-434.

Marche

Cava Giacometti
Cazzella, A. and M. Moscoloni 1994. Il sito stratificato di Cava Giacometti (Arcevia, Ancona) nel quadro degli sviluppi culturali dell'Italia centro-settentrionale dal Neolitico finale all'età del Bronzo. *Quaderni del Museo Archeologico Etnologico di Modena*, I, 89-119.

Santa Maria in Selva

Silvestrini, M., Baglioni, L., Carlini, C., Casciarri, S., Frediani, A., Freguglia, M., Martini, F., Sarti, L. and N. Volante 2002. Il neolitico tardo-finale delle Marche: primi dati su S.Maria in Selva (Treia, Macerata), in). In A. Ferrari and P. Visentini (eds.), *Il declino del mondo neolitico*. Quaderni del Museo Archeologico del Friuli Occidentale, 4, 453-459.

Freguglia, M., Lo Vetro, D. and N. Volante 2005. Santa Maria in Selva di Treia (Macerata): l'area 1, Atti XXXVIII Riunione Scientifica IIPP, 856-860.

Baglioni, L., Laurelli, L. and N. Volante 2005. Santa Maria in Selva di Treia (Macerata): l'area 2. Atti XXXVIII Riunione Scientifica IIPP, 861-868.

Baglioni, L., Casciarri, S., Martini, F. and N. Volante 2005. Santa Maria in Selva di Treia (Macerata): le produzioni dell'area 3. Atti XXXVIII Riunione Scientifica IIPP, 869-875.

Lemorini, C. 2005. Il sito del Neolitico finale di Santa Maria in Selva di Treia (Macerata): analisi funzionale dell'industria litica dell'area 3. Atti XXXVIII Riunione Scientifica IIPP, 876.

Saline di Senigallia

Baglioni, L., Casciarri, S. and F. Martini 2005. Saline di Senigallia (AN), Atti XXXVIII Riunione Scientifica IIPP: 896-900.

Calcinaia, Serra S. Abbondio

Baglioni, L., Casciarri, S. and F. Martini 2005. Calcinaia, Serra S. Abbondio (Pesaro-Urbino). Atti XXXVIII Riunione Scientifica IIPP, 901-906.

Baglioni, L. and S. Casciarri 2005. Il Neolitico Finale di Calcinaia-Serra S. Abbondio nelle Marche: le produzioni ceramiche e litiche della trincea A. In F. Martini (ed.), *Askategi. miscellanea in memoria di Georges Laplace*. Rivista di Scienze Preistoriche, Suppl.1, 509-528.

Pianacci di Genga

Baglioni, L. 2005. Pianacci dei Fossi di Genga (Ancona): l'industria litica dei livelli 1 e 1 A, settore L. Atti XXXVIII Riunione Scientifica IIPP, 907.

Baglioni, L. 2005-2007. Aspetti tecnologici della produzione foliata neo-eneolitica:il caso studio di Pianacci dei Fossi nelle Marche. *Bullettino di Paletnologia Italiana*, 96, 109-128.

Latium

Quadrato di Torre Spaccata

Anzidei, A.P. and G. Carboni (eds.) 1995. L'insediamento preistorico di Quadrato di Torre Spaccata (Roma) e osservazioni su alcuni aspetti tardoneolitici e eneolitici dell'Italia centrale, *Origini* XIX, 55-325.

Lemorini, C., Rossetti, P., Cuomo, G. and M.R. Iovino 1995. I materiali lavorati e le azioni effettuate: la ricostruzione funzionale dell'industria litica del giacimento neo-eneolitico di Quadrato di Torre Spaccata (Roma) mediante l'analisi delle tracce d'uso. *Origini* XIX, 253-276.

Casale di Valleranello
Casali di Porta Medaglia

Anzidei, A.P., Carboni, G. and A. Celant 2002. Il popolamento del territorio di Roma nel Neolitico recente-finale: aspetti culturali ed ambientali. In: A. Ferrari and P. Visentini (eds.), *Il declino del mondo neolitico*. Quaderni del Museo Archeologico del Friuli Occidentale, 4, 473-482.

Abruzzo

Fossacesia

Petrinelli Pannocchia, C. 2003. L'industria litica di Fossacesia (strutture 2-9). Atti XXXVI Riunione Scientifica IIPP, 625-628.

Petrinelli Pannocchia, C. 2005. Analisi dell'industria litica delle strutture 2-9 del villaggio neolitico di Fossacesia (Chieti). In F. Martini (ed.), *Askategi. miscellanea in memoria di Georges Laplace*. Rivista di Scienze Preistoriche, Suppl.1, 421-438.

Settefonti

Radi, G., Castiglioni, E., Formicola, V. and M. Rottoli 1999. Le site du Néolithique recént de Settefonti (Prata d'Ansidonia, L'Aquila). Atti XXIV Congrés Préhistorique de France *Le Néolithique du Nord Ouest Méditerranéen*. Carcassonne, 51-56.

Terenzi, P. 2005. L'industria litica scheggiata del villaggio neolitico di Settefonti (L'Aquila), In F. Martini (ed.), *Askategi. miscellanea in memoria di Georges Laplace*. Rivista di Scienze Preistoriche, Suppl.1, 439-469.

THE BEGINNING OF METALLURGIC PRODUCTION AND THE SOCIOECONOMIC TRANSFORMATIONS OF THE SARDINIAN ENEOLITHIC

Maria Grazia Melis

Dipartimento di Scienze Umanistiche e dell'Antichità,
Università di Sassari (Italy)

Ramona Cappai, Laura Manca and Stefania Piras

Ecole doctorale 355, Université de Provence, CNRS, LAMPEA – UMR 6636 (France)

Corresponding author: mgmelis@uniss.it

Introduction (MGM)

Understanding dynamics of socioeconomic transformations which mark the transition from the Neolithic to the first Copper Age and the development of the Eneolithic in Sardinia requires an analysis of the cultural evolution of the Ozieri phenomenon across its main phases: the Ancient (San Ciriaco), the Middle (Su Tintirriolu), the Recent (Pranu Mutteddu?) and the final (Sub-Ozieri) one (Melis *et al.*, 2007).

Especially between the recent phase and the final one, elements of innovations can be noticed. This will produce a gradual change in the Neolithic substrate even if settling strategies and funerary rituals appears to remain largely unchanged.

In terms of chronology this transition takes place in the centuries between the two halves of the IV millennium (cal. BC). However, radiocarbon data – still too few – show a lack of homogeneity between the North and the South of the island. An important element of innovation, which concerns religious ideology, is represented by the monument of Monte d'Accoddi (Sassari), built during the final phase

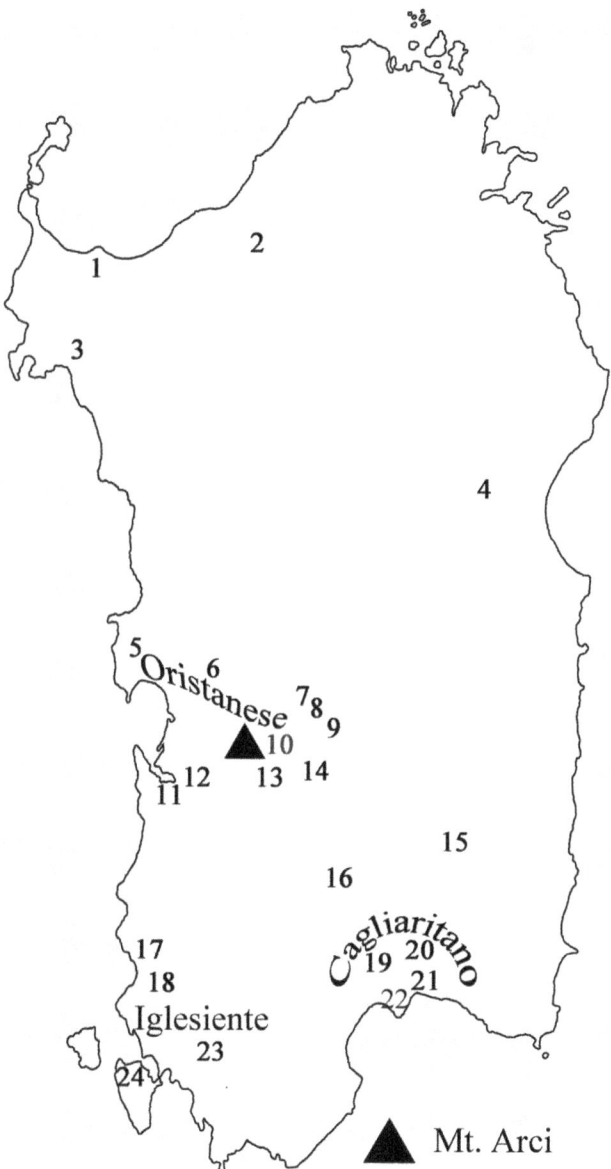

Figure 1: Map of distribution of the sites quoted in the text: 1, Monte d'Accoddi-Sassari; 2, Contraguda-Perfugas; 3, Anghelu Ruiu-Alghero; 4, Biriai-Oliena; 5, Isca Maiori-Riola Sardo; 6, Su Cungiau de Is Fundamentas–Simaxis; 7, Piscina 'e Sali-Laconi; 8, Corte Noa-Laconi; 9, Tamadili-Laconi; 10, Bingia 'e Monti-Gonnostramatza; 11, San Giovanni-Terralba; 12, Murera-Terralba; 13, Enna Pruna-Mogoro; 14, Scaba 'e Arrius-Siddi; 15, Pranu Mutteddu-Goni; 16, Serra Cannigas-Villagreca; 17, San Benedetto-Iglesias; 18, Tanì–Iglesias; 19, Craviole Paderi-Sestu; 20, Is Calitas-Soleminis; 21, Su Coddu/Canelles-Selargius; 22, Capo Sant'Elia-Cagliari; 23, Cungiau Su Tuttui-Piscinas; 24, Campu Scià Maìn /Tupei-Calasetta.

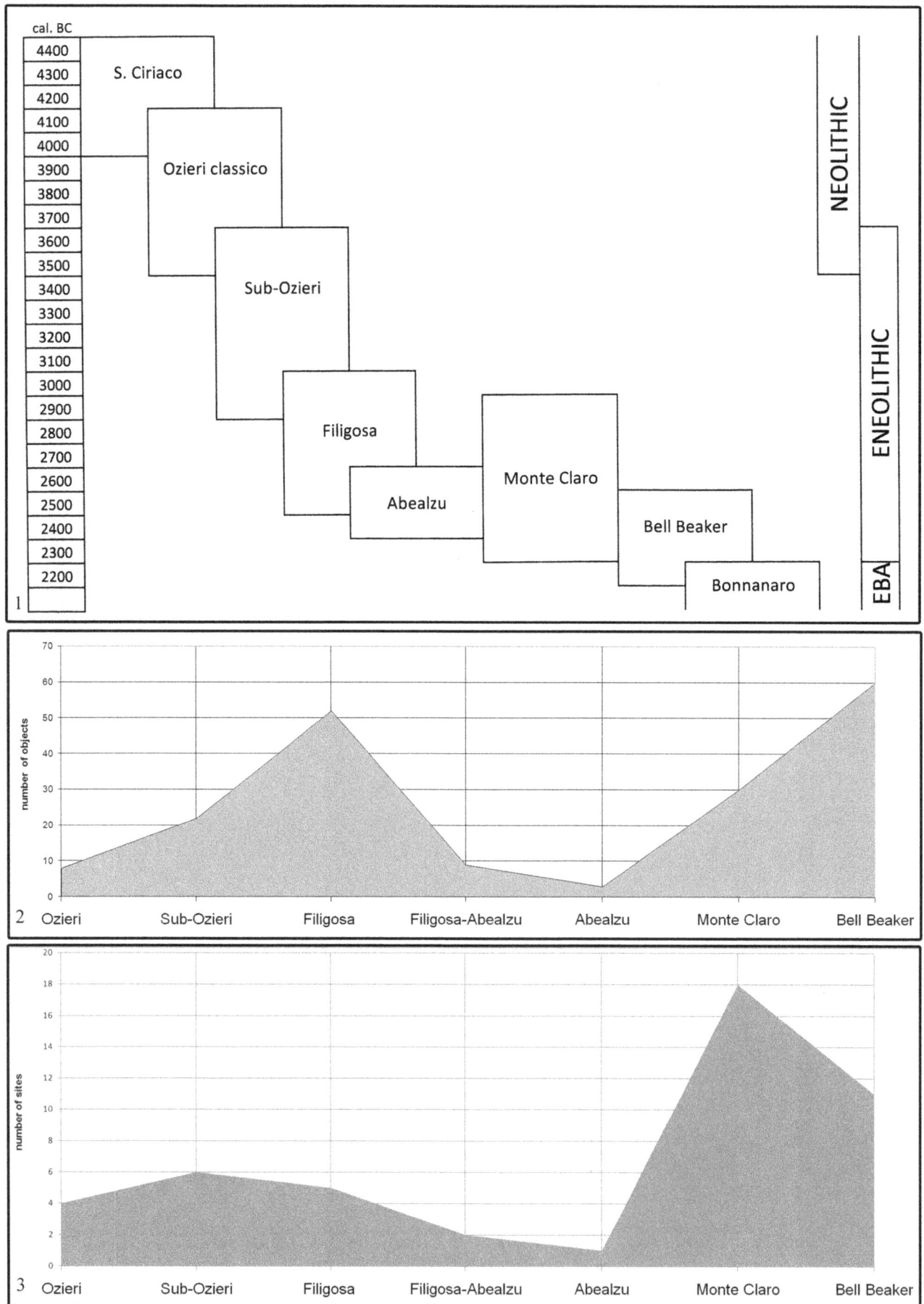

Figure 2: 1, Chronological table of Sardinian Late Neolithic and Eneolithic; 2, diffusion of metal tools in final Neolithic and Eneolithic; 3, distribution of sites with metal finds in Final Neolithic and Eneolithic.

of the Ozieri period on a previous place of worship and dwelling-site referred to the Ancient, Middle and Recent phases of the same culture. A gradual transformation of the socioeconomic organization of late Neolithic and a development of early Eneolithic are perceived thanks to morphological (Melis 2000), technological and functional analyses of craft production, with particular references to interactions among activities related to different raw materials (metals, stones, hard animal materials and pottery). In this sense an important contribution is offered by unpublished data from the excavations in the dwelling-site of Su Coddu-Canelles (Selargius) (Melis *et al.*, 2007). There, an opportunistic production in lithic, ceramic and hard animal materials seems to have been adopted, as the following paragraphs will show in a more detailed way[1].

In the classical Ozieri period the objects are characterized by a low standard of technology, or by rough decorations.

It is amazing, for instance, the disappearance of rich decorations on pottery and, at the same time, the experimentation of a new technique peculiar of the Sub-Ozieri: the vase painting.

It is important to underline that such a change is not tied to the loss of a technical know-how, since it persists in such precious objects as *sub-figulina* pottery or bone beads. Other factors may have determined a change in the rhythms of life, thus reducing the time dedicated to the creation of objects and modifying the ways of raw materials procurement. This "opportunistic" aspect may depend on a new social organization with the rise of a specialized agriculture, as suggested in Selargius by different indirect elements (*eg.*, village expansion, presence of *silos* and *dolia* for the food storage[2]). Does the beginning of metallurgy play any role in such a change?

Metal tools, made up of copper and silver; appear in the recent phase of the Ozieri period (Figure 3; Figure 4, 1). Like necklace beads as those of Pranu Mutteddu (Goni) (Figure 4,1a) and awls, among which an unpublished sample from Monte d'Accoddi (Figure 4, 1b).

The studies on the first metallurgic activities in Sardinia are still rather inadequate: even if on the whole the findings in the last few years have grown richer if compared to the pioneering work by Lo Schiavo (Lo Schiavo 1989)[3]. Metallurgic analyses are scanty, and only a few findings have been subdued to analyses to recognize

	Ozieri	*Sub-Ozieri*	*Filigosa*	*Filigosa-Abealzu*	*Abealzu*	*Monte Claro*	*Bell Beaker*
awls, pins	X	X	X		X	X	X
beads	X	X	X	X		X	X
axes		X	X			X	X
blades		X				X	
dagger blades		X	X			X	X
crucibles		X	X			X	
rings		X	X	X			X
bracelets		X					X
clamps of restoration						X	
tuyeres						X	?
arrowheads							X
necklace (*torques*)							X
mirrors							X

Figure 3: Metal items from the Sardinian Eneolithic

their components, melting techniques and origin of raw materials (Lo Schiavo *et al.*, 2005).Therefore all valuations, conditioned by such a lack of investigation, will exclusively refer to numerical data and to the their original contexts.

First of all, it is important to underline that metal artefacts are equally distributed in living, religious and funerary contexts since their first appearance. During the transition to the Sub-Ozieri no change is recorded except for a strong increase of human presences in living contexts. In the subsequent phases of Eneolithic metal acquires a fundamental role in the composition of funerary equipments, sometimes substituting almost completely lithic artefacts. It seems therefore that during the first metallurgic phases (Ozieri and Sub-Ozieri) metal tools played a role in daily life but only in a second time they acquired the value of status symbols. The data from Su Coddu-Canelles support this hypothesis. That settlement gives us 12 metal objects related to the Sub-Ozieri phase (Figure 4, 2), but one object from an Ozieri hut. The finding of a crucible is the evidence of some kind of metallurgic activity *in situ* (Manunza 2005). Ugas identified some slags in Ozieri and Sub-Ozieri structures, although no chemical analysis have been carried out so far (Ugas *et al.*, 1985),

---

[1] Our team, co-ordinated by the writer and made up of the authors of the present article, intends to apply a functional morpho-technonolgical approach to the study of crafts in the first phases of Sardinian Eneolithic, pointing out the role and the possible interaction among the activities. This is the first time that such a kind of approach is adopted in Sardinian prehistory.

[2] The great importance of agriculture in the early Copper Age compared with the Neolithic is moreover confirmed from anthropological analyses that underline a food diet based more on vegetables than on animal products (Lai *et al.*, 2007).

[3] Neolithic and Eneolithic Sardinian artefacts listed there are about one hundred. Now 234 elements are known - also including objects that

indicate metallurgic activities as crucibles and *tuyères* – 195 of which of a clear cultural attribution. Apart from published findings, some unpublished artefacts from Monte d'Accoddi and Su Coddu are included here. The findings considered are mainly from funerary contexts (77%), less frequently from dwelling-site (14%) and places of worship (9%).

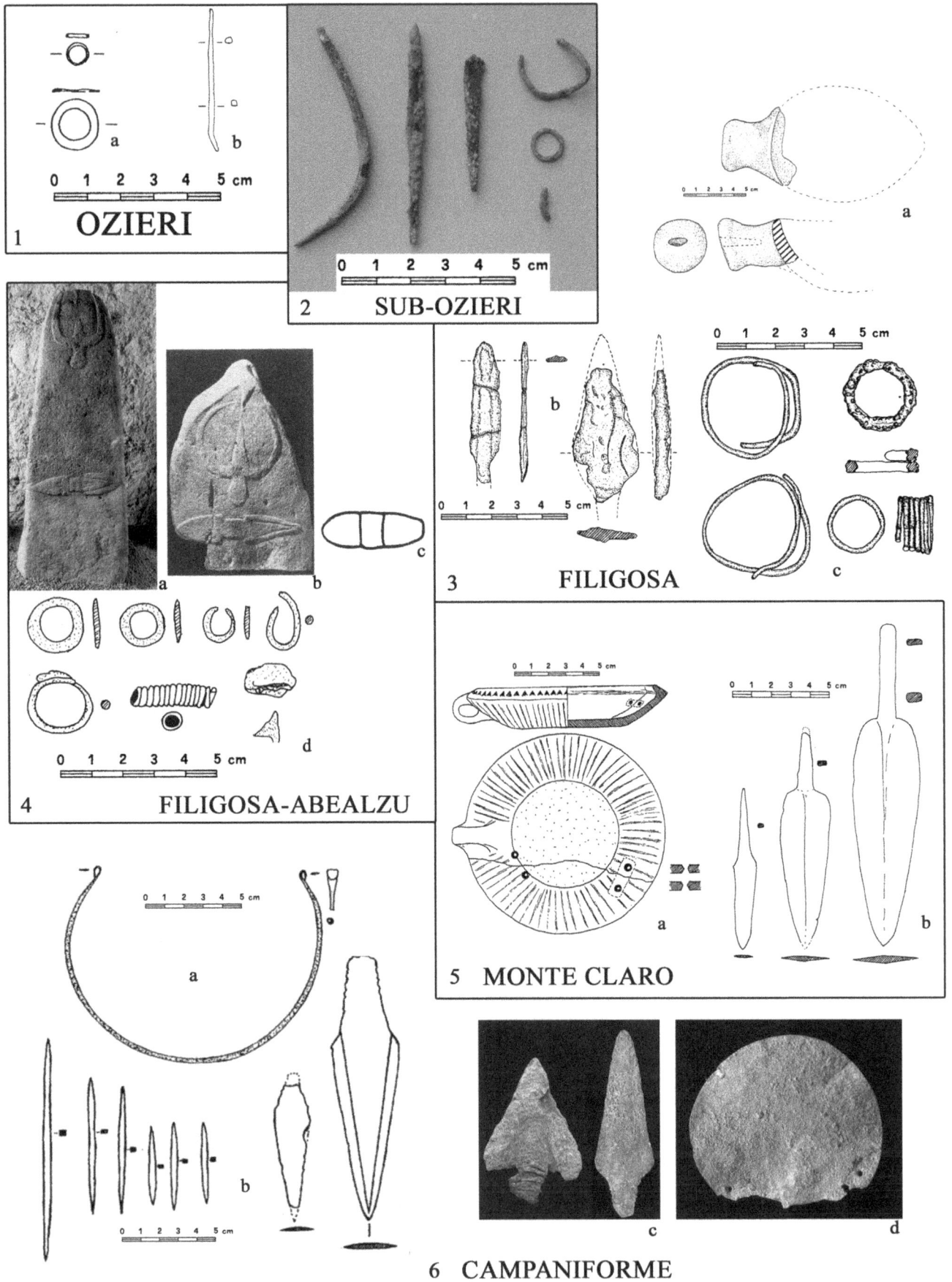

Figure 4: Metal items and indicators of metallurgical activity from different areas of Sardinia.

while the analysis of some slags from the sector of Canelles exclude a correlation metallurgic activities (Melis 2005)[4].

Awls without bone handles, which live together with bone point tools, are the best represented artefacts. Rare in the Filigosa, Abealzu and Monte Claro periods, they become particularly frequent in the Bell Beaker one (31 elements; Figure 4, 6b); in this case, too, they do not replace bone tools similar in shape, which continue to be used.

Daggers - which are present since the Sub-Ozieri, too – from the Filigosa period become an important element in funerary equipments and their symbolic value is mirrored in the so-called *statue menhir* (Figure 4, 4a-b). These ones show us handled daggers whose formal typology is quite homogeneous. Some authors interpret the element opposing the blade as another blade rather than as a knob-handle. As a matter of fact, the two parts are never equal in terms of dimensions and only in one of them a sort of 'V' is sometimes present, representing the flattening of the blade's margin[5]. As for the handle, which is particularly well defined in the *statua menhir* Piscina' e Sali I (Laconi) (Figure 4, 4a), we have no archaeological evidence. A third hypothesis is based on the idea of a portrayal of a metal dagger and a stone mutually opposed and bound to the same central handle (Atzeni 1998). In such a case a lithic object and its metal equivalent may play the same symbolic role. It is important to remember that several *statue menhir* from Laconi are to be connected with the grave of Corte Noa (Laconi), whose equipment is composed of ornamental metal objects, various excellently manufactured obsidian tanged arrowheads and no copper dagger (Figure 4, 4d; Figure 10, 4). On the contrary, in the equipment from Cungiau Su Tuttui (Piscinas) (Figure 4, 3b-c) – quite far from Monte Arci obsidian sources but not from the mineral sources of Iglesiente – five metal daggers are present. In some *statue menhir* there are some important represented objects like the one found in Tamadili (Laconi) where (Figure 4, 4c) – the stylized representation of the dagger evokes the shape of the statue itself. We find a similar example in Lunigiana. At Monte d'Accoddi the considerable presence of metal tools along with that of crucibles (Figure 4, 3a) and other objects referable to a metallurgic activity would suggest that metallurgy was particularly related to the sanctuary's rites, just like spinning and weaving. Moreover, three crucibles represent an important finding considering that in the rest of the island only a crucible from a village Monte Claro and another one from a Sub-Ozieri context are known.

The number of metal tools gradually increases during the Eneolithic, thus witnessing the increase of metallurgic activities that probably, yet slowly; influence the choice of dwelling-sites: as a matter of fact, between the ancient Eneolithic and the recent one a gradual decrease in site distances from the metallogenic areas is registered.

The graph at Figure 2, 2, showing the distribution of objects in each cultural phase, points out a particular concentration in the Filigosa and Bell Beaker ones; nevertheless, in the Filigosa this trend is conditioned by the extraordinary grave findings of Cungiau Su Tuttui (20 artefacts; Figure 4, 3b-c) and Serra Cannigas (Villagreca) (18 artefacts) (Usai 2000; Atzeni 1985). In fact, the curve related to the sites of findings (Figure 2, 3) shows a reduced quantity of locations in the first phase of Eneolithic and a larger distribution in the Monte Claro contexts. In this case the use of lead-clamps for the restoration of ceramic artefacts is recorded (Figure 4, 5a).

During the Bell Beaker phase the metallurgic production reaches its top, even if limited to funerary contexts (Figure 4, 6). Awls and daggers prevail, while the range of *parures* includes small amounts of metal objects as opposed to bone and stone made objects. However, it is exactly in the ornamental category of objects that gold appears for the first time (Figure 4, 6a): a golden and silvery torque comes from the rich grave goods from Bingia 'e Monti (Gonnostramatza) (Atzeni 1998a).

Pottery (SP)

In Sardinian prehistory the beginning of metallurgy is accompanied by a regression phase in the pottery production. These evidences may not necessarily be mutually related, but they both took place during the same period of social, economic and technological mutations. Just like the cycles of different raw materials, the activities put in motion inside a community, too, are not mutually independent: they may share technological advances or similar working procedures. The product and the waste products may be used in another process; the energy needed in one activity is inevitably detracted from another one; the efficacy of a new technology may take the place of an outdated one, etc... for these reasons it is essential to study them as a whole.

As for pottery, in Sardinia no technological study seems to have been systematically applied – not even at a macroscopic level − to any prehistoric artefacts (*i.e.*, on the one hand, the analysis of raw materials, of marks produced during moulding and finishing phases or due to the use of special tools during the working process and, on the other hand, analysis of the relationship between these aspects and vase forms destined to different functions).

As a consequence, little can be said about the activities related to these artefacts and their socio-economic background without ethno-archaeological comparisons, archaeometrical analysis and reproduction experiments.

Nevertheless, using an iconographical repertory from literature as a sort of 'touchstone collection' about it,

---

[4]   Recent experiments have reproduced chemical-physical alterations similar to those found in Selargius by M. R. Manunza: they are compatible with temperatures of about 1000°-1200° and can be referable to metallurgic activities (Manunza *et al.*, 2005-2006).

[5]   That is typical of Bell Beaker daggers.

we could analyse a small amount of ceramics belonging to the Sub-Ozieri phase recorded in the village of Su Coddu-Canelles[6]. The same technological approach has been adopted in the study of hard animal materials and obsidian-made artefacts. Macroscopic detected marks are the result of a variety of moulding technologies, while imperfections and anomalies reveal the work of a 'clumsy' craftsman, incapable or careless about executing properly all the operations needed.

Most artefacts show also surfaces made irregular by particles of degreasing materials, crevices and harshness, all of which witness incomplete or hasty operations. Only a relatively small number of samples let us hypothesize that such an accurate smoothing and polishing of surfaces was intended to give these pieces an undoubted functional and aesthetic value.

Marks related to finishing operations reveal interventions at different times: on a still completely wet paste or at a more or less advanced stage of its drying process; by means of tools with different consistency; after wetting surfaces again or without wetting them at all.

Chromatic differences on surfaces and in sections lead us to hypothesize the use of simple furnaces in the open air, as suggested by burnt vegetable particles in pastes, probable baking accidents and chromatic variations due to post-depositional events. Rare decorated artefacts[7] present a low standard of complexity.

The Sub-Ozieri pottery from Su Coddu-Canelles could be defined an "opportunistic" one because is practical and without any trace of aesthetic care.

It is striking the contrast with the pottery of the previous Ozieri period, which, although lacking detailed technological studies, was extraordinarily refined, rich and various is, as resulting from a high investment in terms of time and work needed. Forms are elegant and geometrically regular and surfaces accurately refined, while homogeneous hues on both inner and outer surfaces denote the mastery of oxidizing and/or reducing baking systems. Decorations obtained by means of a wide variety of techniques and tools characterize a high percentage of artefacts. Sometimes we find decorations covering both the inner and outer faces of a vase, including its bottom, or even spindle whorls and loom-weights. The use of colours (red and yellow ochre, white pastes) for encrustations implies the adoption of a sub-operational sequence based on the procurement of raw materials for pigments, their carrying, transformation and application to obtain a resistant decoration.

All technological, ornamental and symbolic refinements

that affected the Ozieri pottery disappear during the first phases of the Copper Age; on the contrary, a continuity remains in the forms of vases (Melis 2000) (we do not know yet if some kind of continuity affects also the choice of materials and the work sequences) in terms of a gradual evolution: that is why we find it difficult to hypothesize the disappearance of a technical ability (Figure 5).

The results of archaeometrical analysis (mostly published in preliminary studies) concerning the phase of raw material procurement have made it possible to state the dynamics and importance of some regions' exploitation and the relations among the communities that populated them. For instance, the ceramic production from some sites in the Oristanese area is exclusively based on the use of local clays (Bertorino *et al.*, 2000), while a local production also results from the analysis of the ceramics found in Is Calitas (Soleminis) and Su Coddu-Canelles (Cara and Manunza 2005: 34-35), both in the Cagliaritano area.

Nevertheless, the preliminary results from the analysis in course on Su Coddu-Canelles' pieces show a certain variety of pastes and, beside strictly local or sub-local raw materials, other pastes with an allochthonous origin. They characterise some pieces selected on the basis of their anomaly[8].

An "opportunistic" exploitation of local resources is confirmed, but a wider mobility is proved: in such a context ceramic containers witness the introduction of technological or aesthetic 'exotic' products.

Perhaps such mobility is due to the procurement of raw materials (lithic resources, ochre, minerals...) or of other goods whose nature is still unknown.

The ceramics is a material and enduring good whose manufacture is traditionally, as the weaving and the plaiting, a women activity used in a domestic context[9]. The last ones, except for the collection and the preparation of raw materials, can be interrupted and started again (Atzori 1980). On the contrary moulding, finishing and decoring pots require well defined operations at definite stages of the material's plasticity and drying process, so as to obtain the best result. Such carefulness has not been found in the examined repertory.

So why women seem not to have enough time nor care to dedicate to pottery making? What activity absorbs the

---

6 The examined materials come from the stratigraphical excavation of structures 39, 40, 47 and 48 located in the area of Su Coddu-Canelles investigated under the scientific direction of Maria Grazia Melis.

7 The rate of decorated artefacts relating to the first phases of the Copper Age (Sub-Ozieri, Filigosa and Abealzu) is estimated 20% of the whole (Melis 2000: 43).

8 Mineralogical petrographic analyses have been made by Paola Mameli at the Istituto di Scienze Geologico-Minerarie.

9 The reference has to be consider for local ethnographical context: the weaving and the realization of basketry equipments destined to the domestic activities, are a female prerogative (treatment of cereals and flours, food maintenance); the male intervention pertains to other phases like supplying vegetable raw materials (seasonal activity) and the treatment of stronger fibres (olive, marshy reed). The ethnographical and the ethno-archaeological comparisons attribute the ceramic production to women in absence of lathe and furnace (characteristic of the eneolithic ceramic in Sardinia), while men would take over with the advent of such innovations and the overcoming of the domestic dimension of the production.

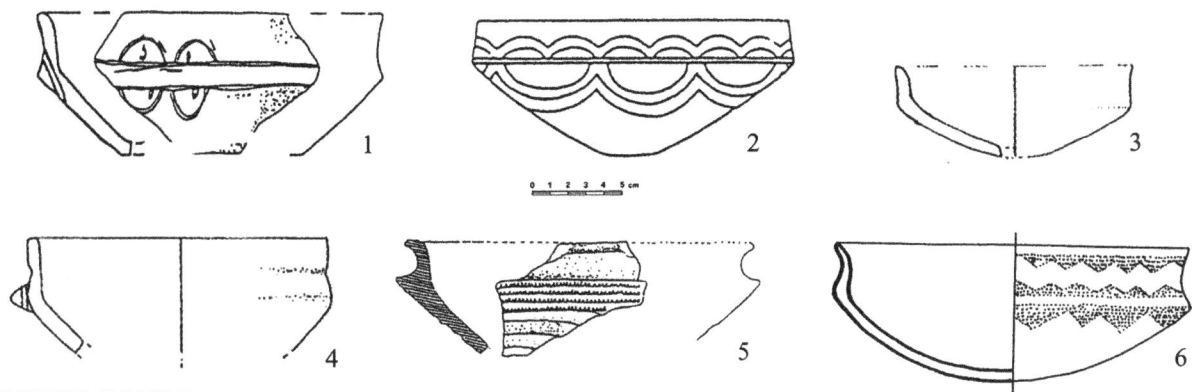

CAREENED BOWLS

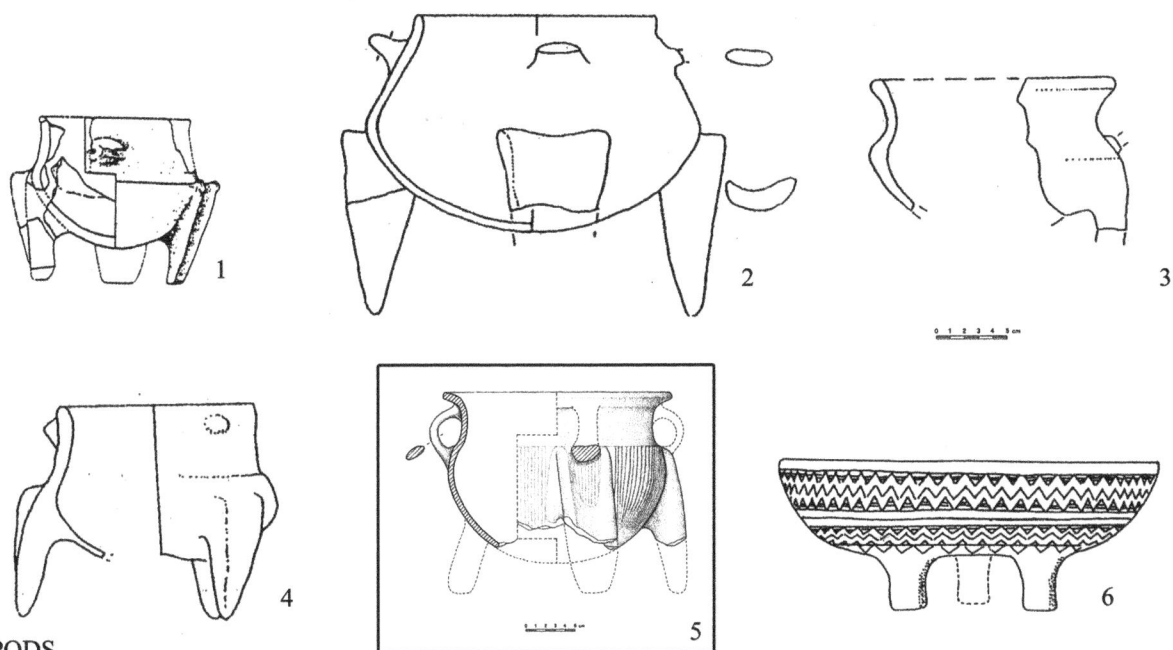

TRIPODS

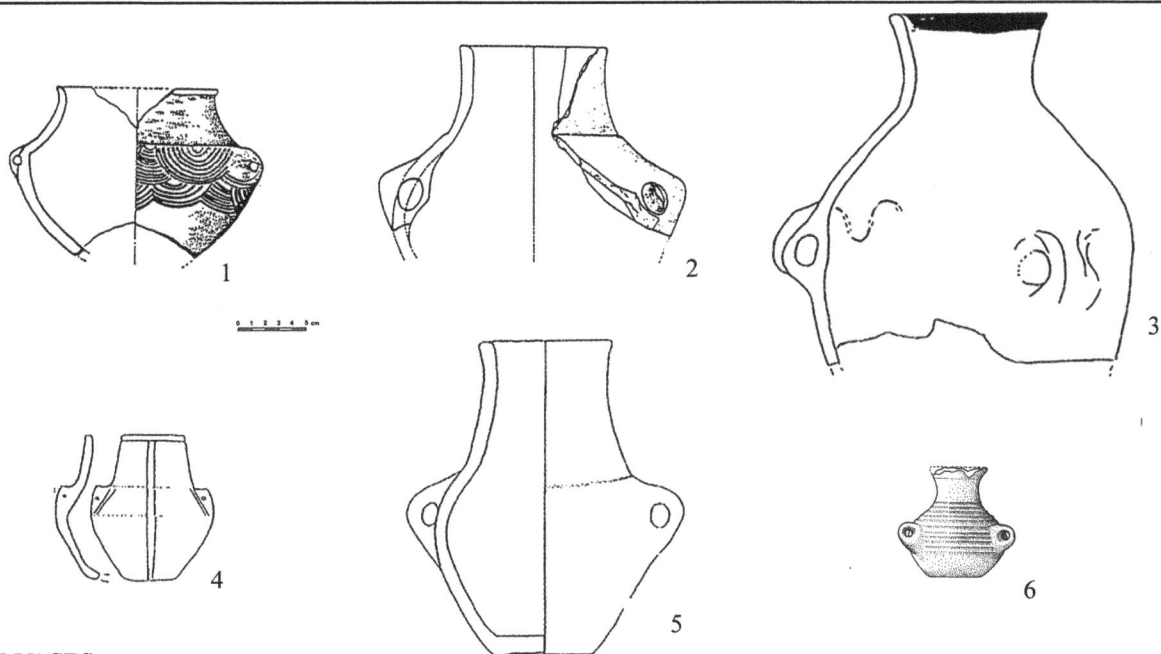

NECK VASES

Figure 5: Some forms from Final Neolithic and Eneolithic Sardinian pottery. Careened Bowls: 1-2, Ozieri; 3, Sub-Ozieri; 4, Filigosa; 5, Monte Claro; 6, Bell Beaker. Tripods: 1, Ozieri; 2, Sub-Ozieri; 3, Filigosa; 4, Abealzu; 5, Monte Claro; 6, Bell Beaker. Neck vases: 1-2, Ozieri; 3, Sub-Ozieri; 4, Filigosa; 5, Filigosa-Abealzu; 6, Monte.

energy amount that had been previously employed in pottery making?

Perhaps farming is the answer, because the increasing development of its technology required the full-time presence of the community. Breeding (sheep breeding for wool, milk and its by-products, cattle breeding for drawing), require a wider communitarian engagement which is inevitably taken off from other sectors not directly connected to subsistence.

During the phases of ceramic moulding and finishing, a macroscopic analysis does not reveal any use of metal tools in the working sequence of pottery. Nevertheless, the site of Su Coddu-Canelles gives us a holder crossed by a hole with a square section (instead of a circular or elliptical one) that may have been produced by means of a metal tool. An occasional circumstance, though.

Finally, some reflections on the phase of baking are necessary. Human ability to control thermal energy (atmosphere and temperature) has been exercised in baking pottery, which implies observing and interpreting reactions, identifying their empirical reasons in relation to raw materials, types of furnaces, fuel and ventilation: in one word, experimenting. All of that experience is supposed to have converged in metallurgical activities, thus giving birth to a parallel branch of development.

Our data refer mainly to some ceramics' baking temperatures (*i.e.*, temperatures developed inside furnaces during the baking process) resulting from archaeometrical analysis and macroscopic observations concerning chromatic aspects of surfaces and sections. Temperatures of Sub-Ozieri ceramics from the Oristanese area reach the 650°; those of Filigosa ceramics from Santu Pedru (Alghero) vary between 650° and 750°, while those of Abealzu ceramics from Monte d'Accoddi between 550° and 750°. Analyses on Filigosa and Abealzu ceramics denote an extreme variability in temperatures and baking conditions, even inside the same batch and the same artefact. However, they are too few and have a geographical sporadic origin to use them to reconstruct an evolution of baking procedures.

Hard animal material (LM)

In Sardinian archaeological literature the artefacts produced in hard animal material have never been subject to an exhaustive analysis. Even if in the last years a step forward has been taken, nowadays the situation is still insufficient. As a matter of fact we lack morpho-typological studies identifying pieces and skeletal parts, employed shapes and working techniques. Furthermore, there are no techno-functional studies pointing out procurement strategies of raw materials, techniques, processes and methods of productions and use. Obviously, a reliable picture of prehistoric Sardinian exploitation of hard animal material will only arise from the filling of these gaps. This study intends to contribute to this field of research. We will try

to identify categories of objects produced in Sardinia from the Ozieri period to the end of the Eneolithic one, when a development in the field of metallurgy is tangible, in order to investigate the possible effects of the introduction of metals. An important contribution will be supplied by the preliminary results of technological analysis of a sample coming from the site of Su Coddu (Selargius), in the area of Canelles[10].

The diffusion of artefacts: the final Neolithic and Eneolithic. Morphological data

The corpus collected takes into consideration the published artefacts produced in Sardinian prehistory, but also some unpublished ones from the dwelling-site of Su Coddu-Canelles. The data collected for each artefact censed affects the site of finding, context and cultural attribution, category, species and skeletal part employed.

The sample results to be statistically representative: it includes artefacts from dwelling-sites and necropolises located either in the North, the Centre and the South of the island (Figure 6).

Only a few findings can be surely attributed to the Ozieri period, since many sites have not been investigated using a stratigraphical method, by means of which chronological separate contexts can be identified. In addition, often no distinction was made between the Ozieri and the Sub-Ozieri levels (Sanna 1999, Figures 6, 6-7, 10; Usai 1987, Figure. 6, 8), since the latter has only recently been recognized. Awls made up of longitudinally rived bones or long bone flakes, small point tools and flat sharp objects are known. The list also includes elements for *parures* (Figure 7) – remarkable the pierced teeth of *canidae* (Foschi Nieddu 1984: 536), or the cylindrical necklace elements whose apex is pierced and pierced valves of *Cerastoderma edule/glaucum*, too (Usai 1990: 70).

The Sub-Ozieri phase includes all the categories described above: only the absence of flat sharp objects is recorded. Awls with a preserved epiphysis, arrowheads and bevelled objects are added. The preliminary results of a recent study concerning the artefacts from Su Coddu-Canelles (Figure 8, 1-5) contribute to characterize this cultural phase in a more detailed way. There, the most employed raw materials come from domestic animals: ovi-caprines and bovines. Among the first ones metapodials are mainly used (Figure 8, 3), even if the employment of tibias is also recorded. While among the second ones metapodials (Figure 7, 4), femurs and radius are exploited. Ornamental objects are made of valves of *Cerastoderma edule/glaucum* whose umboes or top surfaces are pierced (Figure 8, 5) and a small bone bead crosswise pierced.

---

[10] The artefacts taken in to consideration were discovered during the excavations directed by M. G. Melis (2001-2005) in structures 39, 40, 43, 47 and 48 in the area of Canelles (Melis *et al.*, 2007, *ivi* bibliography). The analyzed findings represent the whole of materials from structures 39, 40 and 48; however all finished artefacts from the five structures mentioned above have been taken into consideration.

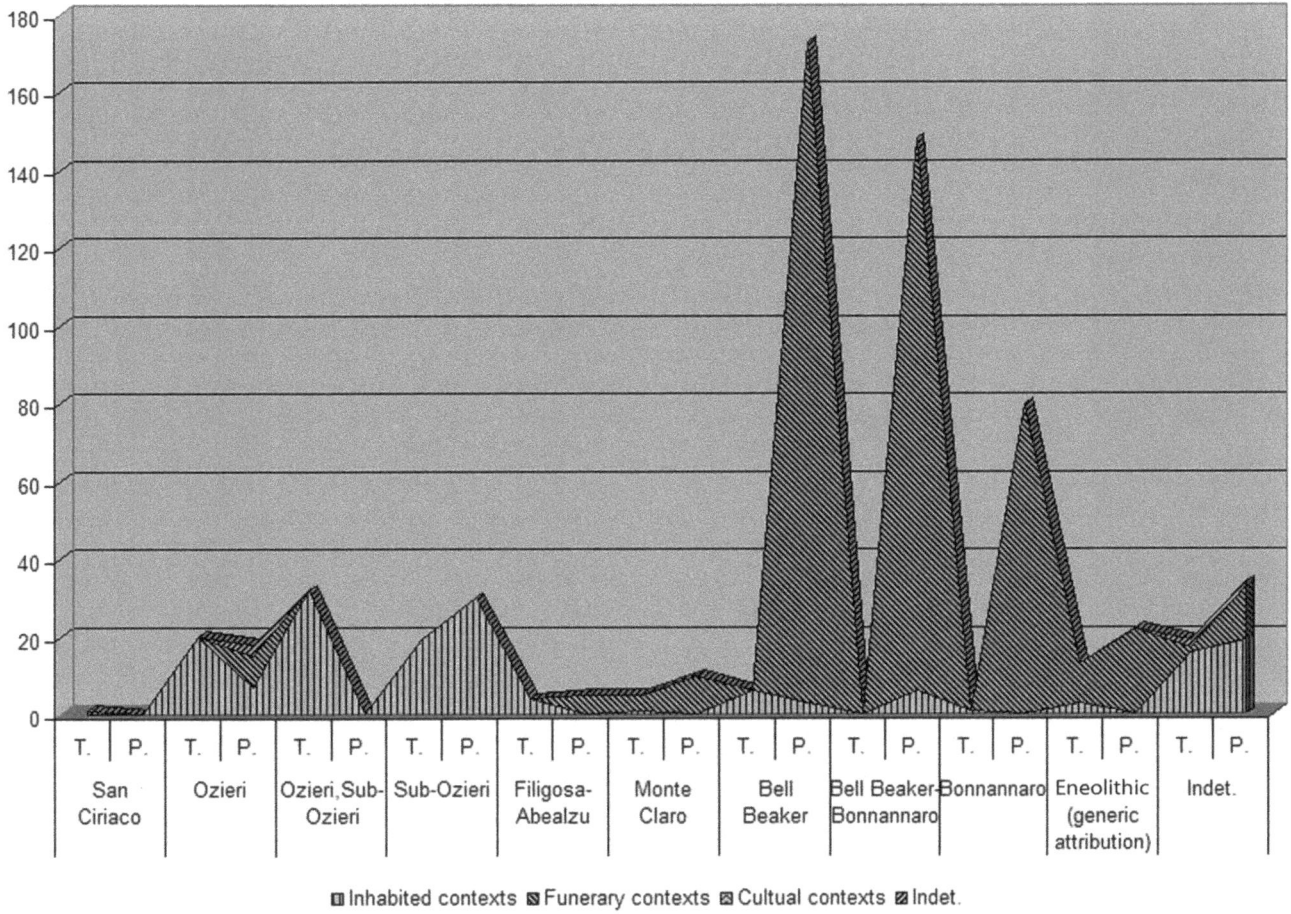

Figure 6: Diffusion of the hard animal material objects from Final Neolithic to Early Bronze age (T. = Tools; P: = Parure).

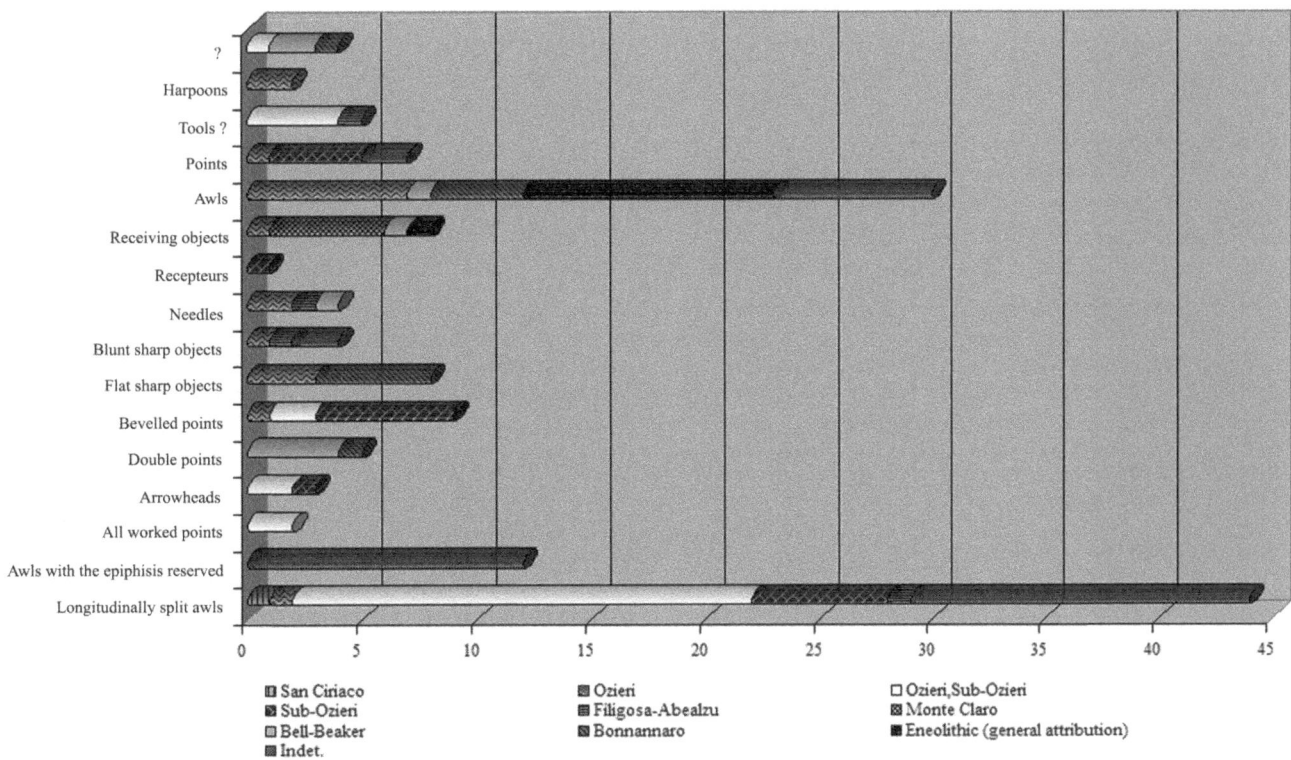

Figure 7: The tools categories of hard animal material and its diffusion from Final Neolithic to Early Bronze age.

Due to scanty evidence we will deal with the artefacts found in Filigosa and Abealzu contexts as a whole. They consist in full-worked points (Atzeni 1985: 30, tav. IV, 1; Melis 2000: 78, fig. 109, n. 35: 332), a point of worked bovine horn (Contu 1997: 309) and a tool made of a long bone diaphysis decorated with several parallel linear engravings[11].

As regards the Monte Claro period, new elements are registered: on one hand the production of tools decreases perceptibly, whereas the range of ornamental objects increases; on the other hand a new category of tools appears: the handled tools. These were employed to make the use of metal points [12] preserved easier and, probably, even the typical long tanged daggers (whose tangs have rectangular sections) had such kinds of handles. An example of that is a handle found at Su Cungiau de Is Fundamentas (Simaxis) (Figure 8,12) (Atzori 1959: tav. III); it was made out of a long bone decorated on both its top surface and edges with crosswise notches. Some portions of epiphyseal long bones, obtained by transversal cuts, have also been found and interpreted as necklace elements. Nevertheless, it is also possible to interpret them – the largest ones, at least – as handles[13].

During the Bell Beaker and the Bonnannaro periods the production of various elaborate elements for *parures* exceeds that of tools, represented by flat sharp objects and double points. The latter may be compared at a morphological level to copper awls. We can assume that: the production of double points, already recorded in ancient Neolithic does not decrease as a consequence of the production of similar copper tools.

In conclusion, a continuum in the employment of hard animal material for the tool production between the Recent Neolithic and the First Copper Age is confirmed, with a light decrease of tools in an advanced phase of the Copper Age.

In the Monte Claro phase the presence of artefacts used to handle copper tools and the decrease in the production of tools in general were observed; on the contrary, a sort of break between the production of Neolithic tradition and the Bell Beaker/Bonnannaro can be noticed, since then the use of hard animal materials has been almost exclusively restricted to the production of objects for *parure* (Figure

8, 13). The decrease in the production of some artefacts does not seem to be due to sudden changes caused by the introduction of metal tools; however, it witnesses a gradual loss of importance in daily activities.

The contribution of technological analysis to the study of hard animal material industry of Su Coddu-Canelles (Selargius)

The current study is important for two reasons: on one hand, it characterizes well the production of the Sub-Ozieri from a morpho-typological point of view, which has been only slightly represented until the present; in fact, materials come from a uncontaminated context. Moreover, this is the first technological analysis on Sardinian bone artefacts. The conditions of raw materials procurement and the techniques adopted to obtain tools will be pointed out by means of an attentive study of the typical technical stigmata and, sometimes, of experimental reproductions. About fifty findings – including waste, blanks and rough-out – and more than 1100 fragments of long bones come from structures 39, 40 and 43 of Su Coddu-Canelles. This study is therefore homogeneous and representative as a whole; moreover, it supplies techno-economical information concerning the community settled at Coddu-Canelles at the beginning of the Copper Age. All finished objects have been found in quite deep stratigraphical unities not upset by subsequent events. As regards the procurement strategies for the tool production, an almost exclusive use of domestic animals' skeletons is registered, except for one artefact obtained from a deer bone. Beads for *parures*, instead, are mainly extracted from *Cerastoderma edule/glaucum* that may have been collected not far from the settlement.

Two kinds of tools are recognizable: pointed objects (Averbouh and Cleyet-Merle 1995: 95)[14] and bevelled tools (Camps-Fabrer *et al.*, 1998: 105), both having a wide range of applications in daily life.

Technologies applied are very simple, either in the phase of *débitage* or in the phase of roughing-out, and are adopted in the same tool when strictly necessary. As a matter of fact, in the phase of roughing-out, scraping is not used to regularize surfaces; it is used more often to eliminate the exceeding portions of raw material, instead: that is the case of points whose shaping has not been finished yet in the phase of *débitage*. A good example is a point used without finishing it and, even if not for a long time, used subsequently. As related to the main theme of the present writing, it is important to underline that no trace of metal tools have been found in any phase of the working sequence.

The whole production of the bone industry of Su Coddu-Canelles seems to have been conceived as a consequence of an urgent need. The production and use of plain tools and a

---

[11] Discovered in the anteroom of the *domu de iana* of Scaba 'e Arrius (Siddi) near a skull (US5) and interpreted as an object destined to adorn the area of the dead.
[12] In any case, in a cave at Capo Sant'Elia, Cagliari, a copper awl was found inside a long hollow bone used as a handle (Figure. 8, 11) (Pinza 1901: tav. III).Unfortunately its chronological references are uncertain even if generally attributed to the Eneolithic.
[13] To support this hypothesis we have to consider that the bone objects for *parure* are more elaborated. Buttons are formed by pierced plates and have long elliptic shapes whose ends are decorated with crosswise engravings; cylindrical artefacts are decorated with transversal notches forming some sort of small globes on surfaces. Instead, roughing-out cylindrical objects are found in contexts where copper points are present, too (copper oxidation traces can be noticed on one of the cylindrical artefacts found in sector *f* of the cave of Tanì (Iglesias) Ferarese Ceruti and Fonzo 1995: 110, 111 e fig. 6, 1-3).

[14] A point whose base is tapered could be associated to double points. These – handled at their base – are used as fishhooks by North-European populations. Nevertheless, ethnographical comparisons remind us that similar objects are used for bird hunting.

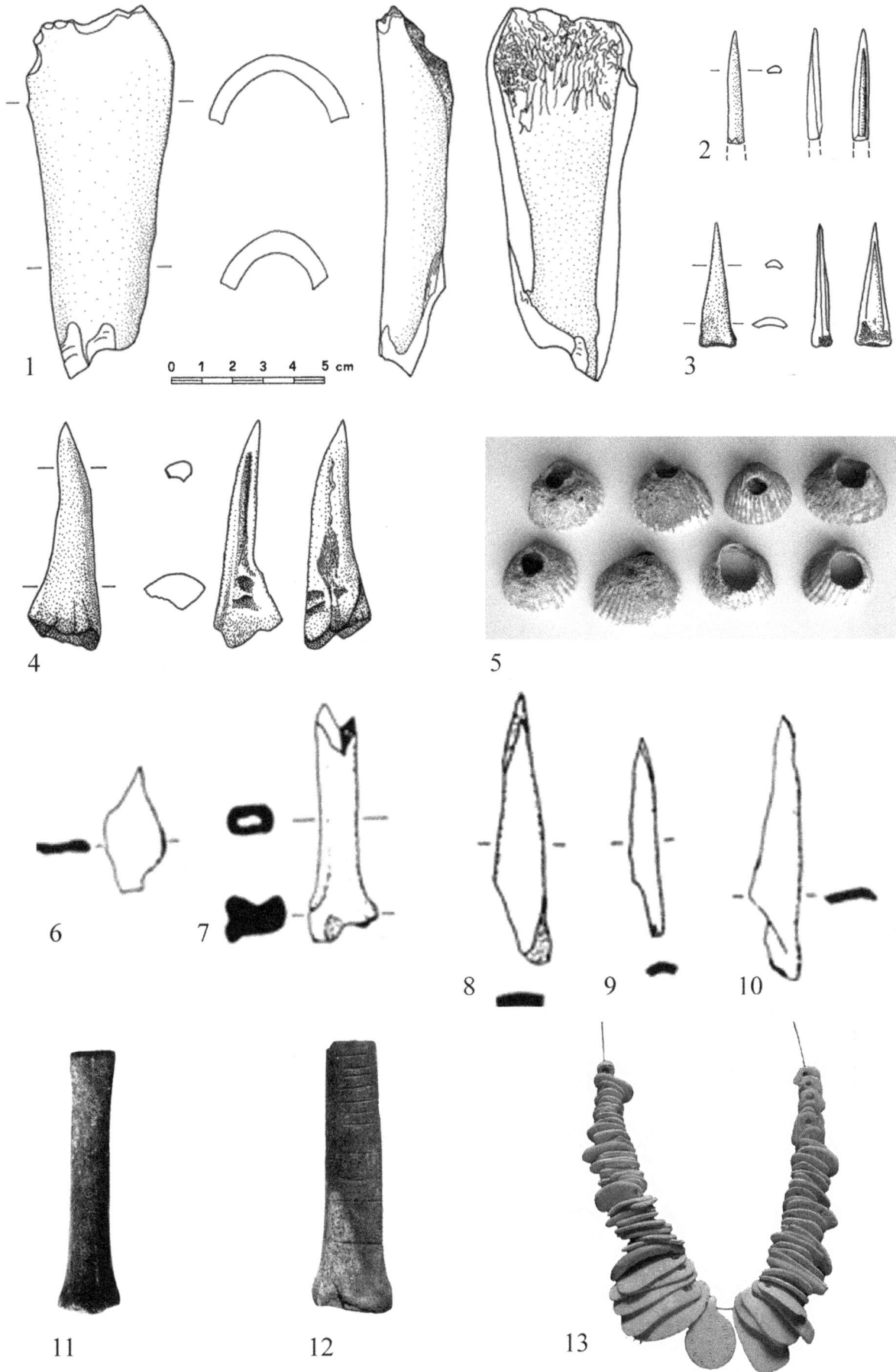

Figure 8: Hard animal material artefacts coming from different areas of Sardinia: 1-5, Su Coddu-Canelles (Selargius); 6, 7, 10, Is Arridelis (Uta); 8, Terramaini (Pirri); 11, Capo Sant'Elia (Cagliari); 12, Su Cungiàu de is Fundamentas (Simaxis); 13, Padru Jossu (Sanluri). The artefacts 11, 12 and 13 are without metric staircase.

sort of easiness are combined; which does not necessarily mean that the technical know-how was reduced: for instance, a bead found in structure 47 is the product of a more elaborated technical process.

Percussion, the most employed technique, often produces fracture surfaces with similar morphologies (i.e. the same angles extent produced by two convergent planes with the same inclination) evoking the idea of a deep knowledge of the effects produced on raw materials as a consequence of the same action. Moreover, the finding of only three blanks confirms the above mentioned hypothesis of a production conceived to meet urgent needs. The contexts where finished objects were found (structures 39 and 47) indicate that these tools were abandoned when still exploitable. The presence of blanks and waste is ascribable to the production of other smaller artefacts as awls or some point objects, since they can all be referred to the working of long bones of medium sized animals'. Unfortunately, no information is available concerning the working process needed to produce bevel-point tool, since we only have finished objects.

To sum up, we can assert that:

- in the site of Su Coddu-Canelles the transformation of hard animal material takes place by means of very simple techniques and quite unelaborated schemes of transformation;
- raw materials come almost exclusively from domestic animals;
- artefacts are characteristically basic for the absolute lack of finishing;
- almost all the artefacts have not been used for a long time and, in almost all cases, have been abandoned when still exploitable[15].

All these elements contribute to make us think of a rather standardized utilitarian production whose aim was not that of creating a certain amount of objects at a time or of saving raw material, which was always at hand.

The disuse of some tools, abandoned when still exploitable, might be explained with a need for more effectiveness as compared to new productive requirements or a functional inadequacy in comparison with other artefacts (metal tools?).

As for the daily activities performed at that time we can add that the production we found at Su Coddu-Canelles shows a community that, with certain exceptions and despite its elaborated technological heritage (deducible from the analysis of the bead *chaîne opératoire*), is no more interested in the creation of artefacts and ornamental objects.

A flourishing breeding of ovi-caprines was the basis for the procurement of meat, leather and hard animal material.

A broadening of studies concerning complete analyses of the findings from Su Coddu-Canelles will enrich the picture described above.

In order to understand if any direct contact has ever occurred between bone artefacts and similar products in copper (awls, small pointed artefacts and double points) we are planning an experimental program intended to reproduce and verify, comparing their chemical-physical properties and functionalities during the execution of different activities, the possible analogies and differences between the two.

Lithic industry (RC)

The stone tool assemblages

If every technique is the result of mental schemes tied to collective ways of seeing and realizing things, changes in techniques involve transformations in the inside dynamics of the society.

These changes can be regulated by different laws, according to the technological choices of a community within a limited series of alternatives: there exist, in fact, various ways of making the same thing and only the differences in the technical procedures can underline social differences. Because of this, the study of technology is important for the understanding of the phenomena of social interaction.

In Sardinia the lack in technological studies does not allow us to trace a general picture of the characteristics of the lithic industry based on the same methodology because of the nonhomogeneity of the data, on the one hand, and of the disparity in the contexts recovered on the other.

It has often been underlined that the moment of passage between Neolithic and Copper Age is characterized by a "contraction" in the use of stone tools evident by its smaller presence in archaeological data, a consequence of the increasing interest in metal, of which we start to have the first attestations. But in this moment the small recovery of metal tools and of manufactures directly involves in its production can't endorse this hypothesis. Moreover recent technological studies (Cappai 2007) on lithic industry have underlined a difference in human behaviours toward a raw material already known and used rather than a total abandonment of the same one.

In the Late Neolithic of Sardinia, despite the presence of few studies - and those that exist are directed above all to surface materials[16]- it is clear that the presence of "amorphous" flakes, with cortex or with big dimensions, not

---

[15] A tool, for example, with surfaces covered with evident scraping striae and a small blunt. This may imply that it was re-sharpened or even abandoned just a short time after being produced.

[16] It is to underline the analysis of Campu Scià Main – Tupei by G. Vacca in Melis and Vacca 2000 and the analysis of Craviole Paderi – Sestu in Cappai 2003, 2006.

yet exhausted cores, unworked blocks, and great quantity of supports and tools documents a strongly delimited tendency in the use of raw material.

The application of *chaîne opératoire* studies shows that the strategies of acquisition, production and consumption were for the greatest part organized in similar ways. The raw material reached the site in form of a raw block, here it was subject to cortex removal, roughing out the shape and then knapped for the production of usually standardized blanks. Examples in this sense are the materials collected in the site of Craviole Paderi (Sestu) (Figure 9, 1-14). The analysis of the lithic industry, exclusively in obsidian, has allowed the identification and chronological isolation in particular, of two different technological systems. While the second can be connected with the First Copper Age thanks to the comparisons with the materials of Su Coddu-Canelles (Selargius) the first one is a typical Final Neolithic assemblage: the presence of raw blocks and a still exploitable core of obsidian, a high percentage of cortex removal flakes, roughing out and resharpening elements, testifies that the raw material reached the site in the form of completely raw or semi-worked blocks.

The economy of the *débitage* includes the production of standardized blanks as blades or flakes, directly *in situ* in a diversity of ways. There is, in fact, great attention in the production of rather regular and straight blades and bladelets with a planed or facetted butt, probably extracted by indirect percussion, without excluding the possibility that pressure-flaking was used, as testified to by a small overshoot blade. For the first phases of reduction, however, the use of the direct percussion with a tender hammerstone can be supposed, indicated by the recovery of a partially exploited blades core and the presence in the assemblage, of characteristic elements like the absence of a well defined contour in the posterior part of the butt, that isolates the impact point, characteristic bulb scars and the presence of very closed hackles (Pelegrin 2000) These supports have rarely been transformed completely.

No difference in the economy of raw material is found. Obsidian, represented by different type naked eyed distinguished by their visual characteristics, is the exclusive raw material present. This fact is fairly tied up by factors such as the criteria of surface's collection and by the geographical position of the site that allowed easy access to the sources. Obsidian in this chronological moment is, in fact, prevalent in the sites of the centre-south of Sardinia while flint is more regularly exploited to the north, both accompanied by secondary raw materials such as jasper, chalcedony or quartz.

In general, in the raw material economy, a tendency to realize short blades, arrowheads and bifacial shaping pieces, as well as non-formal tools (with irregular side removals) in obsidian – because this is the most important material – is evident, while the flint, whether of local origin or not, is preferred for the creation of longer blades often retouched

along the whole profile. This dichotomy is present in several contexts of Sardinia and is often accompanied by the presence of waste of obsidian knapping, while the flint is generally recovered in the form of finished objects.

Because of the absence of technological studies of these contexts[17] today it is impossible to speak about an economy of the raw material; nevertheless some signs can direct us in the formulation of a hypothesis. In the site of Contraguda (Perfugas), recent studies have put in evidence the presence of a probable workshop of good quality local flint for the production of long blades with high technical investment (Figure 10, 1), not found in the site, whose cores were reused for shorter blades and flakes then used in the same site. It is for this reason that the authors hypothesize that these could be produced on the place to be carried to other sites. The know-how and the technical investment testify, without any doubt, to the extrafunctional and prestigious value granted to these blades (Costa and Pelegrin 2004). In fact, it is often in burial contexts that we can find this kind of objects. Known in all phases of Sardinian Eneolithic, tombs are key contexts to understand variation in the management of the raw material or the presence of symbolic objects. More direct examples of this are the grave II of the necropolis of St. Benedetto in Iglesias (Figure 10, 3) and the necropolis of Pranu Mutteddu in Goni (Figure 10, 2). Chronologically related to the later phases of the Ozieri phenomenon, the two graves have returned burial equipments rich in symbolic elements[18]. The management of the raw material underlines the presence of long blades made exclusively in flint (the size is compatible with the blade product in Contraguda[19]) instead the arrowheads, partly very accurate, are exclusively in obsidian. A small dagger blade and a stiletto, made in flint too, emphasize the particular character of the equipment.

This panorama changes with the first phases of the Copper Age. In close relationship to elements of continuity, new ones witness a deep change. In Su Coddu-Canelles settlement the preliminary study done on the material coming from some structures (Cappai 2007) and belonging exclusively to the Sub-Ozieri phase, shows deep changes in the production and management of raw material in comparison to the preceding period (Figure 9, 15-23). The assemblage is made up exclusively of obsidian (only a fragment of bladelet is in flint) but at the moment we cannot establish the original state in which it was introduced to the

---

[17] It is important to highlight that this study is based on the analysis of several contexts by the author (Cappai 2003, 2003a, 2006, 2007; Cappai and Melis 2006) and of a new lecture of just published assemblages.

[18] For the tomb of San Benedetto refer to Atzeni 2001. Besides the presence of blades and arrowheads, there are a small polished axe in green stone, a polish-pierced knob and a necklace made of stone bead. For the tomb of Pranu Mutteddu refer to Atzeni and Cocco 1989. The rich equipment in knapped lithic industry is accompanied to a polish-pierced knob, a small polished axe and two circular elements in silver, among the first metal items present in Sardinian Neolithic.

[19] This affirmation intends to underline only the probable existence of a circuit related to this kind of objects made in a particular raw material and not the connection between the settlement of Contaguda and the two necropolises. In fact, there are not technological studies and provenance analyses on the raw material that can establish s direct relationship.

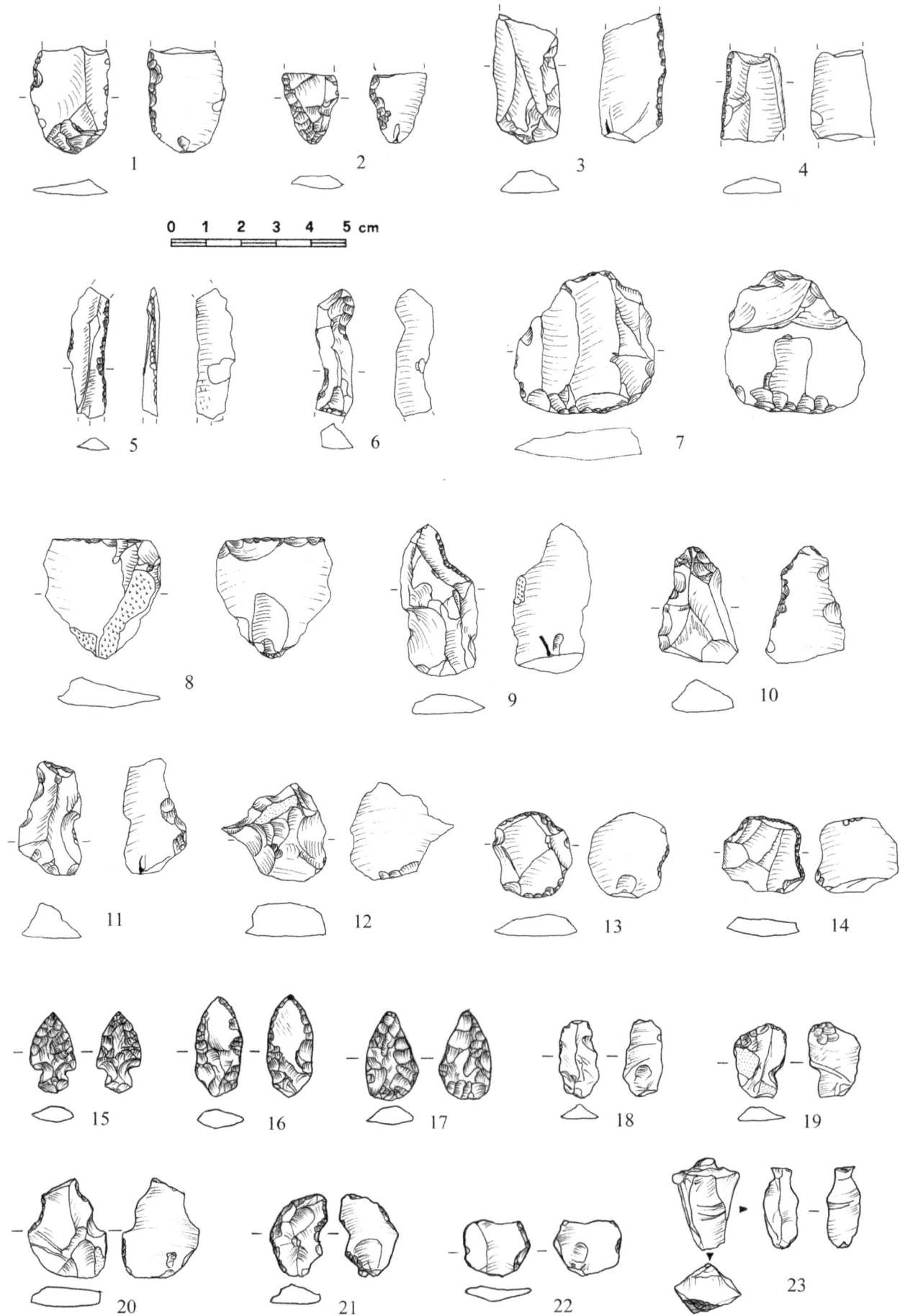

Figure 9: Some examples of lithic tools: 1-14, Craviole Paderi (Sestu); 15-23, Su Coddu-Canelles (Selargius).

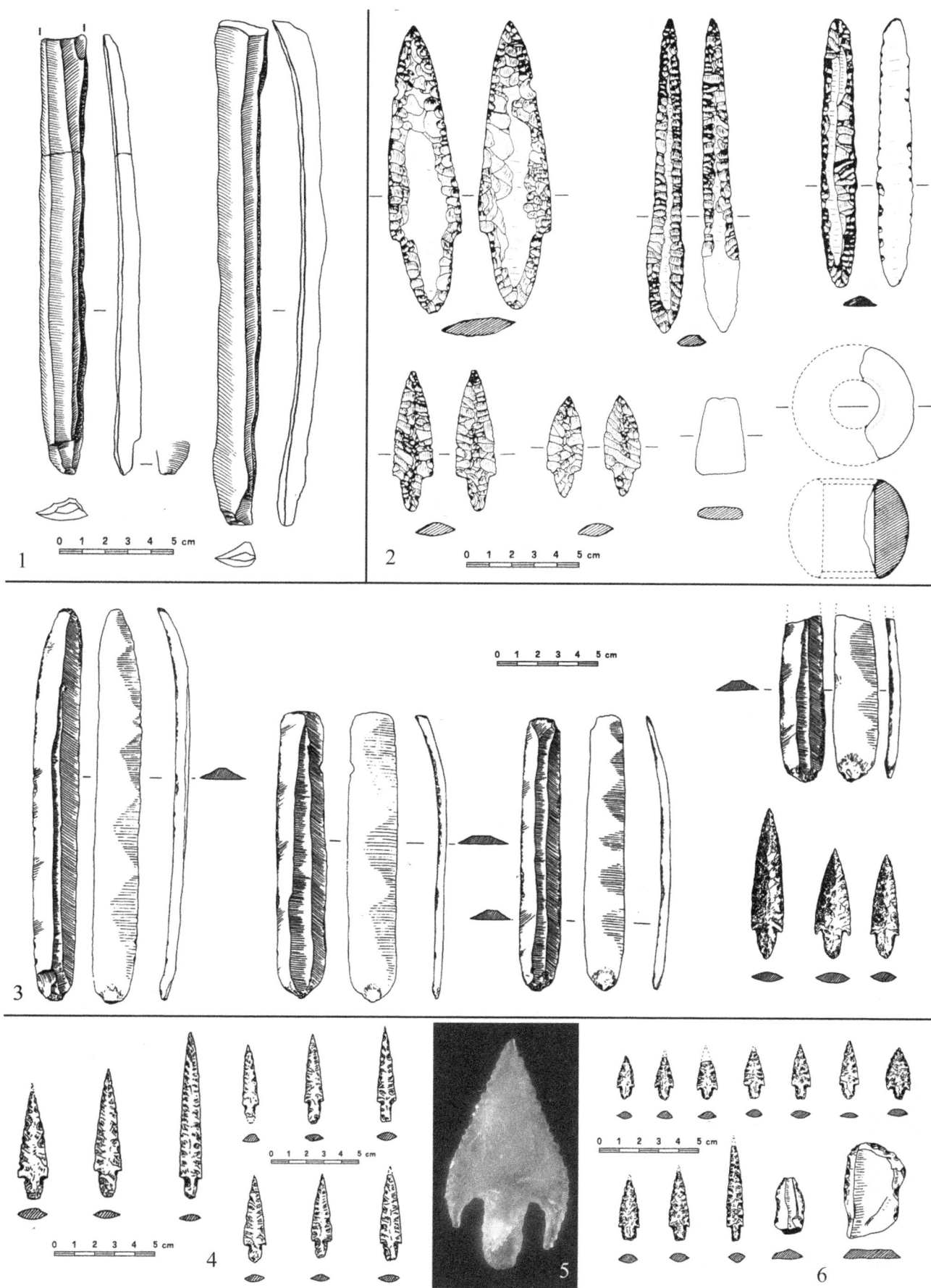

Figure 10: 1, Long blades from Contraguda (Perfugas); Examples of lithic tools from some funerary equipments: 2, Pranu Mutteddu (Goni); 3, San Benedetto (Iglesias); 4, Corte Noa (Laconi); 5, Anghelu Ruiu (Alghero); 6, Serra Cannigas (Villagreca).

site. Even if not predominant, materials with cortex are also present. The cortex can be like a thin patina, a middle or well-developed cortex depending on the distance from the mother rock. These characteristics, in fact, are indicative of different procurement strategies that provide for the collection directly either from the primary and sub-primary source or from secondary sources (Luglié, Le Bourdonnec *et al.*, 2006). On one side the reduced dimensions of the tool assemblage and the presence of residual cores, bring us a strategic picture in which also the small nodules of raw material are exploited for knapping, from the other we cannot exclude the possibility of exploitation of greater blocks for the presence of shaping out and resharpening flakes of great dimensions. In the first case, experimental tests (unpublished personal researches) show that reduction sequences can also be realized from small rounded pebbles or more geometric nodules of small dimensions (7-10 cm). In this case, the nearly complete removal of the well-developed cortex is necessary.

The laminar production is poor and without high technical investment, only indirect elements confirm a well-organized *débitage* that contrasts with a non-negligible percentage of bipolar percussion artefacts that first was recognized in Sardinian assemblage (Cappai, in press). This regards the production of flake or laminar flake like in small residual cores as is testified to by a refitting effected in the structure n. 40 (Figure 9, 23). The most representative blanks are, in all cases, small and middle-sized flakes, a lot of which are fragmentary.

The present stigmata on the other materials, testify reduction probably by direct percussion. The knapping accidents point out, in fact, the non homogeneous propagation of the ripples because of the way of striking with either excessive or a lack of force, causing big irregularities on the inferior surface. A single "Siret" accident directs us on this thesis, while the various fractures in the distal ends, can also be caused by the use of other techniques. The presence of hinge terminations can be partly tied to the use of a hammerstone too light in weight or to the fact that it was applied with too little force (Sollberger 1994). Also the toolkit is non-formalized, often of microlithic size and on flakes that have preserved their original morphology. Only the arrowheads and the ogives introduce an intrusive and sometimes covering retouch that implies a high technical investment.

In general this behaviour is observed also in other contemporary contexts as Craviole Paderi from a side, that, not away from Su Coddu-Canelles, proposes some similar schemes: tendency to microlithisme with the exploitation of raw material up to its total exhaustion, reutilisation, low technical investment and Isca Maiori (Riola Sardo) (Depalmas 1989) from the other whose industry appears rather poor, essentially on flakes with prevalence of scrapers while the presence of bifacial shaping pieces is very limited. For the Oristanese area, the phase of passage is for instance documented from the lithic industry of Murera (Terralba) (Cossu 1996) On one side this site notices a

substantial maintenance of the formal typologies found in the Recent Neolithic site of S. Giovanni (Terralba) (Cossu 1996). In comparison with this moment the flint toolkits almost entirely disappear and above all the blades, present instead to S. Giovanni, while the size of the production is rather reduced. The curated shaping is entirely reserved to arrowhead and bifacial pieces.

As a rule, the Sub-Ozieri lithic industry underlined a tendency in the microlithisme: low investment in the production of formal blanks, with very simple and non-formalised *chaînes opératoires* for the production of expedient tools that introduces the bipolar technique too. Only the arrowhead production detaches notably from the whole industry betraying *savoir-faire* and a very important investment of labor.

Bearing in mind the attestations of Contraguda and the data presented here, is there an implication for the presence of specialized groups in the production of formal tools? Or as the ethnographic data show, was it decided to devote more time to the manufacture of stone tools destined to more important and intense activity and that implied a greater risk of failure (Khun 1989; Tomka 2001)? Or, is the opportunistic attitude for the production of the greatest part of the tool assemblage, tied up to the dispersion of the work force in other activities (experimentation in metal production)?

These are working hypotheses for a future research because of the lack of current data and the nonhomogeneity of the contexts of recovery. In fact, except for Monte Claro, for the most part of the Copper Age the greatest part of the material originates from funeral contexts. The grave of Scaba 'e Arriu (Siddi) (Ragucci and Usai 1999), for example, has returned material belonging to various phases, from the Recent Neolithic to the Late Copper Age. With the heavy presence of obsidian arrowheads related to various periods that is typical of Sardinian funeral equipment, it is evident the presence of flint blades in the sub-Ozieri, Filigosa, Abealzu and Monte Claro levels. This witnesses the prolongation in the use of flint blades that are just disappeared by the contemporary villages (the data are related to Sub-Ozieri and Monte Claro phases). In this case we cannot exclude the practise of reutilisation of more ancient objects just present in the grave, which would explain the absence both in other funeral contexts and in the inhabited areas.

The data coming from the funeral contexts, unfortunately often fragmentary, do nothing but guarantee the hypotheses already advanced: great care in the production of particular tools as arrowheads, often in obsidian or in other specific materials such as that in chalcedony of Anghelu Ruiu (Alghero)(Demartis 1998) (Figure 10, 5), use or reutilisation of flint also for the following periods to testify a long symbolic duration of these materials. The «crisis» of stone tools could have been before, then reinterpreted in terms of a change in the approach to the procurement strategies, with a consequent opportunistic attitude in the production of common use tools, low technical investment for the production of the greatest part of the tools assemblage, with the exception of the arrowhead. The use wear analysis, still in progress, can also clarify the operation and functionality of such stone tools [20]. Also during the Monte Claro, in fact, this scheme is repeated. Despite of the few presence of stone tool, both in Biriai (Oliena) (Castaldi 2000), but even more in Enna Pruna (Mogoro) (Lilliu and Ferrarese Ceruti 1960.), the same mental schemes are represented but the lithic industry continues to be attested to and used both in dwelling and funeral contexts.

Lithic industry and metal: opposition or coexistence?

Even if we cannot talk about a "crisis" of the lithic industry, it is clear that metal has played an important role in the communities of the Copper Age even if, according to the current studies, we cannot think of a real utilisation for the production of common use tools, both for their absence in housing contexts, and for the difficulty to imagine such a rapid acquisition of its technology.

While its value is attested by the recoveries in burial contexts, it is difficult to understand if has had a certain part in the realization of manufactured articles in stone. In the first case, to quote only two examples belonging to different moments of Sardinia Eneolithic[21], the grave A of Serra Cannigas (Villagreca) (Atzeni 1985) (Figure 10, 6) or that of Corte Noa (Laconi) (Atzeni 1988) (Figure 10, 4). The two graves have returned a rich equipment of obsidian arrowheads: about ten those preserved during the excavation for the first one and about forty the second. These are connected in a different way with metal tools: with dagger blades and ringlets in copper the grave of Serra Cannigas, ringlets and spirals in copper and silver the second. Are we perhaps in a moment in which the symbolism addressed again by objects connected with certain activities is somehow interchangeable? To the same moment of Serra Cannigas, belongs the grave 2 of Cungiau Su Tuttui (Piscinas) (Usai 2000) in which this proportion is completely distorted: the assemblage contains above all metal tools, both related to objects of ornament but above all tied to copper weapons. The reduced dimensions and the very thin size of this kind of objects betray an extreme brittleness for practical use, a characteristic that has led the author to think that they could be the miniatures exemplary of those of common use, nevertheless not testify. It is remarkable, however, to underline the absolute lack, in this context, of lithic arrowheads while the lithic industry is represented only by some flakes. Can we hypothesize therefore a continuation of the same ritual practices in which old symbols overlap to new materials? Is this the meaning of metal and stone daggers mutually opposed and bound to the same central handle portrayed in *statue menhir*

(Atzeni 1998)? What is evident is the absolute call of the little dagger of this grave to arrowheads attested in other contexts. And how can we consider the copper arrowheads in Bell Beaker graves that follow the similar stone models?

Regarding the use of the metal in the production of lithic tools, the data we have are still fragmentary. If for Italy the use of the copper retoucher has been hypothesized for the ligurian arrowheads (Leonardi and Arnoboldi 1998) and the bifacial shaping pieces of the Lessinia (Isotta and Longo 2004), in Sardinia it has been verified only for the case of Contraguda. In fact, the formal typology of the blades produced in an area of the site and its stigmata, justify the use of pressure with a lever fitted with a copper point. It is not possible, for the moment, to understand if other flint blades were realized with this technique or if this technique was reserved only to this material/product. Nevertheless if the theory of the authors is correct, their extrafunctional value would connect them with ritual-funeral contexts, a privileged domain for which these products were very probably destined. Some reflections can be made on the arrowheads. While some do not reveal any indications in this sense and in general it is not always simple to distinguish pressure using horn from that of a copper point, from the experimental data furnished by other contexts (Isotta and Longo 2004) there are Sardinian examples that recall that type of laminar retouch, regulate, oblique and continuous. These are in particular some stone tool discoveries in the site of Murera, in a context of the First Copper Age, realized in obsidian and quite surprisingly, of some precedent recovered in that of S. Giovanni realized both in obsidian and in chalcedony. Non-valuables include the arrowheads, also very accurate, recovered from Serra Cannigas and Corte Noa.

In light of these data, unfortunately fragmentary because of the state of the research and type of approach, it is possible to return to the question of precise technological choices in the so-called "crisis" of the lithic industry that, at least for the whole Copper Age remains, like that of the hard animals' tools, an important material of the prehistoric communities. Metal, perhaps still in a phase of experimentation and not fully approved for daily activities on which the maintenance of a community depended, does not enter this circuit if not as a symbolic element and very probably only in the production of certain typologies of tools.

Conclusions (MGM)

The whole picture of Sardinian Copper Age[22] as it emerges from the present study is characterized by a slow and gradual evolution through which the Late Neolithic socioeconomic organization undergoes transformations due to technological innovations and new trends in religious ideology.

A discrepancy takes place between a sort of technological

---

[20]   Cappai PhD in progress.
[21]   We can chronologically insert the grave of Serra Cannigas in the Filigosa phase and the grave of Corte Noa in the later phase of Filigosa-Abealzu; Melis 2000.

---

[22]   For further deepening on the various aspects of the Sardinian Eneolithic see Melis 2000, *ivi* previous bibliography.

decadence (or, more precisely, a restriction in the use of a certain technical know-how) in the pottery field, in the lithic and hard animal material industries on the one hand and, on the other hand, the development of metallurgy, building techniques (Monte d'Accoddi) and agriculture.

As for metallurgy it is important to underline as the awareness of the intrinsic and symbolic power of metal comes only later: in fact, the first evidences in the Ozieri and in the Sub-Ozieri phases mainly refer to dwelling-sites. From the Filigosa phase metal tools become important elements in funerary equipments. The symbolic value of daggers is particularly emphasized by their being represented in the *statue menhir*.

In the transition from the Ozieri to the Sub-Ozieri ceramics show elements of tradition in terms of morphology, but also a radical change in decoration which betrays a reduction in manufacturing times. On the contrary, archaeometric analyses on materials from Su Coddu-Canelles suggest us that the origin of raw materials is to be localized in places halfway between the Cagliaritano area and the mining basin of Iglesiente and/or other sources of raw materials used in the site.

At the beginning of the Copper Age the hard animal industry does not seem to have been conditioned by the introduction of metals, which, in fact, do not play any role in the different phases of the *chaînes opératoires* of Su Coddu-Canelles' tools.

As for the lithic industry, in the Sub-Ozieri a reduction in the presence of lithic elements is always perceivable; this shows changes in the procurement of raw materials. In the Filigosa an equivalent presence of prestigious stone and metal tools in funeral equipments has got a double meaning: on the one hand the use of lithic tools is going to fade, whereas more sophisticated technologies are almost always adopted for grave precious goods; on the other hand, the use of metal is going to spread out starting from a phase of experimentation and use in domestic contexts up to the introduction in funeral equipments as status symbols in the non-egalitarian societies of late Eneolithic. But, the composition of equipments seems to be still conditioned by the distance variable from the sources of raw materials procurement. This explains the absence of copper daggers and the presence of beautiful obsidian arrowheads in Corte Noa, not very far from Monte Arci; besides, it also explains the absence of obsidian arrowheads and the presence of 5 daggers out of 20 metal tools from the grave of Piscinas in the mining basin of Iglesiente. Even if a large amount of data referring to the Bell Beaker documents a widespread use of metal tools, it offers a vision exclusively limited to burial contexts. Nevertheless, the higher homogeneity of documents dating back to the Monte Claro allows us to evoke a picture of late Eneolithic where metallurgy does not seem to be exclusively destined to symbolic and funeral aims, since it takes part in daily life interacting with

other activities, as witnessed by metal cramps for restoring ceramics and bone handles for metal objects.

In conclusion it can be affirmed that the introduction of metal doesn't seem to have conditioned the handicraft production and the socioeconomic order in the first phases of the Eneolithic. The transformations, at the end of the Neolithic, seem to be tied up rather to an increasing development of agriculture. Beginning from the Filigosa burial equipments metal assumes an undoubted symbolic role and with the Monte Claro we assist to the spread of metal and its interaction with the other handicraft activities is confirmed.

The continuation of researches through integrated analyses (*e.g.*, archaeozoological, archaeobotanic ones, etc.) on the materials from su Coddu-Canelles will offer more exhaustive answers to a subject that is still strongly conditioned by an heterogeneity of data and a lack of radiocarbon dating.

References

Atzeni, E. 1985. Tombe eneolitiche nel Cagliaritano. In *Studi in onore di Giovanni Lilliu per il suo settantesimo compleanno*, Cagliari, Stef Press, 11-49.

Atzeni, E. 1988. Tombe megalitiche di Laconi. *Rassegna di Archeologia* 7, 526-527.

Atzeni, E. 1998. Le statue-menhir di Piscina 'e Sali, Laconi – Sardegna. In Actes du 2me colloque International sur la statuaire mégalithique, Saint-Pons-de-Thomières.

Atzeni, E. 1998. La tomba ipogeica di Bingia 'e Monti. In F. Nicolis and E. Mottes (eds.), *Simbolo ed enigma. Il bicchiere campaniforme e l'Italia nella preistoria europea del III millennio a.C.*, 254-260 Trento.

Atzeni, E. 2001. La necropoli di cultura "Ozieri" a San Benedetto di Iglesias (Ca). In E. Atzeni, L. Alba, and G. Canino (eds.), *La collezione Pistis-Corsi e il patrimonio archeologico del comune di Iglesias. Mostra archeologica e fotografica*, 25-29. Iglesias.

Atzeni, E. and D. Cocco 1989. Nota sulla necropoli megalitica di Pranu Mutteddu-Goni. In L. Dettori Campus (ed.), *La cultura di Ozieri. Problematiche e nuove acquisizioni*, Atti del I Convegno di Studio, (Ozieri, gennaio 1986 – Aprile 1987). Ozieri, Il Torchietto, 201-216.

Atzori, G. 1959. Stazioni Prenuragiche e Nuragiche di Simaxis (Oristano). *Studi Sardi* XVI, 267- 299.

Atzori, M. 1980. Artigianato tradizionale della Sardegna. L'intreccio. Corbule e canestri di Sinnai. *Quaderni Demologici* 2. L'asfodelo, Sassari.

Averbouh, A. and J.-J. Cleyet-Merle 1995. Hameçons. In H. Camps-Fabrer (ed.), *Éléments barbelés et apparentés. Fiches typologiques de l'industrie osseuse préhistorique*. Cahier VII, 83-99. Treignes, Cedarc Press.

Bertorino, G., Lugliè, C., Marchi, M. and R. Melis T. 2000. Insediamenti preistorici nell'Oristanese (Sardegna centro-occidentale). Primo approccio archeometrico. In Atti della VI Giornata "*Le scienze della terra e l'archeometria"*, Este (Padova), 26-27 febbraio 1999, 185-192.

Camps-Fabrer, H. (ed.) 1998. *Biseaux et tranchants*. Fiches typologiques de l'industrie osseuse préhistorique, Cahier VIII. Treignes, Cedarc Press.

Cappai, R. 2003. L'uso dell'ossidiana nell'insediamento di Craviole Paderi – Sestu (Cagliari). Tesi di Laurea (under the direction of M. G. Melis and M. Mussi).

Cappai, R. 2003. L'industria litica nella prima Età del Rame. In AA.VV., *L'ossidiana del Monte Arci nel Mediterraneo. La ricerca archeologica e la salvaguardia del paesaggio per lo sviluppo delle zone interne della Sardegna*, 86. Pau (Or) 28-30 novembre 2003.

Cappai, R. 2006. L'industria litica in ossidiana nell'insediamento prenuragico di Craviole Paderi – Sestu (CA). In O. Soddu and P. Mulè (eds.), *Sestu. Storia di un territorio dalla preistoria al periodo post-medievale*, 57-70. Dolianova.

Cappai, R. 2007. Su Coddu – Canelles (Selargius – Cagliari): i manufatti litici. Tesi di Specializzazione, (under the direction of A. Cazzella).

Cappai, R. and M.G. Melis 2006. L'ossidiana delle tombe 3 e 32 di Ispiluncas – Sedilo (Or). Un approccio tecnologico. In AA.VV., *L'ossidiana del Monte Arci nel Mediterraneo. Le vie dell'ossidiana nel Mediterraneo ed in Europa. Tecnologia delle risorse e identità culturale nella preistoria*, Atti del 4° Convegno di Studi, 61-72. Pau 17 Dicembre 2005, Comune di Pau.

Cappai, R. in press. Tecnologia della produzione nella sacca 40 del sito di Su Coddu-Canelles, Selargius (CA). In Atti XLIII Riunione Scientifica IIPP *L'età del Rame in Italia*, 26-29 novembre 2008. Bologna.

Cara, S. and M. R. Manunza 2005. Indagine archeometrica su materiali ceramici provenienti dagli scavi archeologici nel territorio di Soleminis. In M.R. Manunza (ed.), *Cuccuru Cresia Arta. Indagini archeologiche a Soleminis*, 273-288.

Castaldi, E. 1999. *Sa Sedda de Biriai: (Oliena, Nuoro, Sardegna): villaggio d'altura con santuario megalitico di cultura Monte Claro*. Quasar, Roma.

Contu, E. 1992. Nuove anticipazioni sui dati stratigrafici dei vecchi scavi di Monte d'Accoddi. Informatica e stratigrafia. In AA.VV., *Monte d'Accoddi, 10 anni di nuovi scavi*, Istituto Italiano Archeologia Sperimentale, 21-36. Genova.

Contu, E. 2001. Monte d'Accoddi tra esegesi, confronti e cronologie. Qualche nuova considerazione. In G. Serreli and D. Vacca (eds.), *Aspetti del megalitismo preistorico*, 59-66, O.C. Sa Corona Arrubia, Dolianova.

Cossu, T. 1996. Le stazioni preistoriche di San Giovanni e Murera - Terralba (Or). *Studi Sardi* XXX, 21-64.

Costa, A.M. 1990. L'insediamento preistorico di Monte Luna. In D. Salvi and L. Usai (eds.), *Museo Sa Domu Nosta*, 69-73. Cagliari, Stef press.

Costa, L. J. and J. Pelegrin 2004. Une production des grandes lames par pression à la fin du Néolithique, dans le nord de la Sardaigne (Contraguda, Perfugas). *Bulletin de la Société Préhistorique Française* 101, 4, 867-873.

Depalmas, A. 1989. Il materiale preistorico di Isca Maiori

nella collezione Falchi di Oristano. *Studi Sardi* XXVIII, 37-59.

Demartis, G. M. 1998. La cultura del vaso campaniforme ad Anghelu Ruiu – Alghero (Sassari). In F. Nicolis and E. Mottes (eds.), *Simbolo ed enigma. Il bicchiere campaniforme e l'Italia nella preistoria europea del III millennio a.C.*, 281-285.Trento.

Ferrarese Ceruti, M. L. and O. Fonzo 1995. Nuovi elementi dalla Grotta funeraria di Tanì (Carbonia). In *Carbonia e il Sulcis. Archeologia e territorio*, 95-115. Oristano, S'Alvure press.

Foschi Nieddu, A. 1984. I risultati degli scavi 1981 nella necropoli prenuragica di Serra Crabiles, Sennori (Sassari). In *The Deya conference of Preistory*, British Archaeological Reports, International Series, 229, 533-555.

Guilaine, J. 1991. Roquemengarde et les débuts de la métallurgie en France méditerranéenne. In AA.VV., *Découverte du métal*, 279-294. Picard,.

Kuhn, S. L. 1989. Hunter-Gather Foraging Organization and Strategies of Artifact Replacement and Discard. In D.S. Amick and R. P. Mauldin (eds.), *Experiments in Lithic Technology*, British Archaeological Reports, International Series 528., 33-48, Oxford.

Isotta, L.C. and L. Longo 2004, Tecno-tipologia dei foliati ottenuti con ritocco seriale su supporto laminare, il caso dei Monti Lessini (Verona) e il loro inquadramento culturale nei contesti eneolitici dell'Italia Settentrionale. *Padusa: Bollettino del Centro Polesano di Studi Archeologici ed Etnografici* 40, 51-72.

Lai, L., Tykot, R. H., Beckett, J. F., Floris, R., Alba, L., Forresu, R., Goddard, E. and D. Hollander J. 2007. Nutrizione ed economia nella Sardegna del sudovest tra il Neolitico recente e il Medioevo: primi dati isotopici. Summary. In XVII Congresso degli antropologi italiani. Cagliari.

Leonardi, G. and S. Arnoboldi 1998. Approccio analitico allo studio delle cuspidi di freccia liguri. In A. Del Lucchese and R. Maggi (eds.) *Dal diaspro al bronzo*, La Spezia, 48-52.

Lilliu, G. and M. L Ferrarese Ceruti. 1960. La facies nuragica di Monte Claro: sepolcri di Monte Claro e Sa Duchessa-Cagliari e villaggi di Enna Pruna e Su Guventu-Mogoro. *Studi Sardi* XVI, 1958-59. 3-266.

Lo Schiavo, F. 1989. Le origini della metallurgia ed il problema della metallurgia nella cultura di Ozieri. In L. Dettori Campus ed., *La cultura di Ozieri. Problematiche e nuove acquisizioni*. Atti del I Convegno di Studio, (Ozieri, gennaio 1986 - Aprile 1987), Ozieri, Il Torchietto, 279-293.

Lo Schiavo, F., Giumlia-Mair, A., Sanna, U. and R. Valera (eds.) 2005. *Archaeometallurgy in Sardinia: from the origins to the early iron Age*. Montagnac, Mergoil.

Luglié, C., Le Bourdonnec, F.–X., Poupeau, G., Bohn, M., Meloni, S., Oddone, M. and G. Tanda 2006. A map of the Monte Arci (Sardinia Island, Western Mediterranean) obsidian primary to secondary sources. Implications for Neolithic provenance studies, 995-1003.

<http://france.elsevier.com/direct/PALEVO/> [Accessed 30 January, 2007]

Manca, L. 2007. L'industria in materia dura animale dell'insediamento eneolitico di Su Coddu (Selargius, Cagliari): analisi morfo-tecnologica, Tesi di Laurea, (under the direction of M. G. Melis and M. Zedda).

Manunza, M. R. 2005. La vita a Soleminis nella Preistoria. In M.R. Manunza (ed.), *Cuccuru Cresia Arta. Indagini archeologiche a Soleminis*, 27-39. Grafica del Parteolla, Dolianova.

Manunza, M. R., Lecca, A., Atzeni, C. and L. Massidda 2005-2006. Lo scavo del lotto Deiana nel villaggio di Su Coddu – Selargius (CA). *Quaderni della Soprintendenza archeologica per le province di Cagliari e Oristano* 22-I, 3-17.

Melis, M.G. 2000. *L'età del Rame in Sardegna: origine ed evoluzione degli aspetti autoctoni*, Soter, Villanova Monteleone.

Melis, M. G. 2005. Nuovi dati dall'insediamento preistorico di Su Coddu-Canelles (Selargius, Cagliari). In 6th Conference on Italian Archaeology, *Communities and Settlements from the Bronze Age to the Early Medieval Period*, Neolithic session, Groningen (Netherlands), April 15-17 (2003), British Archaeological Reports, International Series, 1452 (II), 554-560 Oxford. BAR Publishing.

Melis, M. G. 2007. La Sardegna e le sue relazioni con la Corsica tra la fine del Neolitico e l'età del Rame. In 128e Congrès des sociétés historiques et scientifiques, *Relations, échanges et coopération en Méditerranée*, 253-263, Bastia, du 14 au 21 avril 2003. CTHS, Paris,.

Melis, M. G., Mameli, P. and S. Piras 2006. Aspetti tecnologici e morfologici della ceramica eneolitica. Nuovi dati dall'insediamento di Su Coddu-Canelles (Selargius, Cagliari). In Atti XXXIX Riunione Scientifica IIPP *Materie prime e scambi nella Preistoria italiana*, II, 1232-1235. Firenze.

Melis, M. G., Quarta, G., Calcagnile, L. and M. D'Elia 2007. L'inizio dell'età del Rame in Sardegna. Nuovi contributi cronologici. *Rivista di Scienze Preistoriche* LVII, 185-200.

Melis, M.G. and G. Vacca 2003. Insediamento e ambiente naturale nella preistoria e nella protostoria del territorio di Calasetta (Ca). *Studi Sardi* XXXIII, 7-34.

Pelegrin, J. 2000 Les technique de débitage laminaire au Tardiglaciaire: critères de diagnosi et quelques réflexions. In *L'Europe Centrale et Septentrionale au Tardiglaciaire*, Table-Ronde de Nemours, 13-16 mai 1997, Mémoire du Musée de Préhistoire d'Ile de France, 7, 73-86.

Pinza, G. 1901. Monumenti primitivi della Sardegna. In *Sardinia, Monumenti Antichi 1901-1944*, Carlo Delfino, reprinted in 2002.

Piras, S. 2007. Analisi tecnologica della ceramica eneolitica di Su Coddu – Canelles (Selargius, Cagliari). Materiali dalle strutture 39, 40, 47 e 48. Tesi di Laurea, (under the direction of M. G. Melis, and P. Mameli).

Ragucci, G. and E. Usai 1999. Nuovi contributi allo studio della Marmilla prenuragica: La tomba di Scaba 'e Arriu (CA). *Studi Sardi* XXXI, 111-196.

Sanna, R. 1989. Il villaggio di Is Arridelis- Uta. In L. Dettori Campus (ed.), *La cultura di Ozieri. Problematiche e nuove acquisizioni*. Atti del I Convegno di Studio, (Ozieri, gennaio 1986 - Aprile 1987), Ozieri, Il Torchietto, 231-238.

Sollberger, J.B. 1994. Hinge fracture mechanics. *Lithic Tecnology* 19,1, 17-20.

Tanda, G., Minghetti, G., Mura, A., Pittui, G., Oggiano, G., Meloni, S. and M. Oddone 1988. *Sull'origine della cultura Ozieri: contributo di indagini chimico-fisiche*, Antichità Sarde, Studi e Ricerche, Sassari.

Tanda, G., Mura, A. and G. Pittui 1999. Analisi archeometriche su ceramiche di cultura Abealzu e Filigosa. In *Archeologia delle isole del Mediterraneo Occidentale*, Antichità Sarde, Studi e Ricerche, 5, 161-180. Sassari.

Tomka, S. A. 2001. An Ethnoarchaeological Study of Tool Design and Selection in an Andean Agro-Pastoral Context. http://links.jstor.org/sici?sici=10456635%28200112%2912%3A4%3C395%3AAESOTD%3E2.0. CO%3B2-S> [Accessed 10 January, 2007].

Ugas, G., Lai, G. and L. Usai 1985. L'insediamento prenuragico di Su Coddu (Selargius-Ca) Notizia preliminare sulle campagne di scavo 1981-1984. *Nuovo Bullettino Archeologico Sardo* 2, 7-40.

Usai, L. 1987. Il villaggio di età eneolitica di Terramaini presso Pirri (Cagliari). In *Preistoria d'Italia alla luce della ultime scoperte. Vol. IV, Atti del IV Convegno Nazionale di Preistoria e Protostoria*, Pescia (8-9 Dicembre 1984), 175-192.

USAI, L. 2000. La tomba n. 2 di "Cungiau su Tuttui" in territorio di Piscinas (Cagliari). Nota preliminare. In Congresso Internazionale *L'ipogeismo nel Mediterraneo. Origini, sviluppo, quadri culturali*, Sassari - Oristano, 23-28 maggio 1994, 875-886. Sassari.

Usai, L., Demartis, G. M. and G. Ugas 1998. Catalogo. In F. Nicolis and E. Mottes (eds.), *Simbolo ed enigma. Il bicchiere campaniforme e l'Italia nella preistoria europea del III millennio a.C.*, 296-330. Trento.

Buried Without Metal:
The Role of Lithic Kit in Chalcolithic Funerary Contexts of the Marche Region (Central Italy)

Cristina Lemorini

Dipartimento di Scienze dell'Antichità
Sapienza Università di Roma, Rome (Italy)
E-mail: cristina.lemorini@uniroma1.it

Introduction

At the beginning of the diffusion of metal work in the Italian peninsula, during the Copper Age (3600-2200 cal. BC), the most conspicuous metal tool findings are related to burials.

Daggers, halberds, axes, awls, personal ornaments are present throughout the peninsula, showing regional differences in their morphology and in their combination with other elements of the burial kit. In particular, daggers and halberds are associated to male individuals and represent the power expressed by the symbols of war and violence, the first signs of martial display that will flourish later, during Bronze Age.

A recent "pilot" analysis of the technological and use traces on a sample of axes, daggers and halberds from various Chalcholitic contexts, mostly burials, from Central Italy (Dolfini, 2010) suggest that axes were "mundane" tools whereas daggers, and especially halberds, were related to the burial ritual .

The metal axes found in the burials often show traces of woodworking or other activities. These tools probably represented an important part of the work kit belonging to the individual when he was alive, and were buried with him, even though sometimes still functional.

On the other hand, daggers and halberds probably were associated only with the ritual sphere, since the former show light traces that would fit well with a ritual activity of skinning and butchering, and the latter are totally unused, suggesting a purely symbolic function of martial display and power.

If metal plays an important role in Chalcolithic burial rituals in the Italian peninsula, what is the role of the lithic tools in the same contexts?

Chipped arrowheads, knives, daggers, polished axes, mace-heads etc … are present, with or without metal objects, in the majority of the funerary contexts of the Italian peninsula (Cocchi Genik 1996).

The presence of lithic artefacts in burials is a distinguishing phenomenon of Chalcolithic times and suggests that, as in the case of metal, these items played an important role in funerary rituals like never before.

What was the meaning of these objects?

If present in the same funerary context, did lithic and metal tools have the same purpose and the same value, or did they symbolize a different social status?

Furthermore, did lithic artefacts substitute metal objects in those communities that did not have an easy access to metal in terms of raw material sources, technology or exchanges?

Or is it possible that, even though access to metals and metallurgy techniques was possible, metal was deliberately omitted from the ritual, while a prominent role was given to other types of artefacts such as lithic or bone objects or pottery?

The use-wear analysis results of the lithic kit from Central Italian funerary contexts discussed in this article intends to provide a useful contribution to this topic.

These funerary contexts are all located in a restricted geographical area, behind the Adriatic coast of the Marche Region (Figure 1). They belong to an early phase of the Chalcolithic period (IV millennium cal. BC.; in several cases the first half of the millennium). The funerary kit of each burial consists of various types of chipped stones, polished stones, osseous objects and pottery, but they lack metal objects, even though the presence of metal is attested in the region (Palmieri and Cazzella 1999).

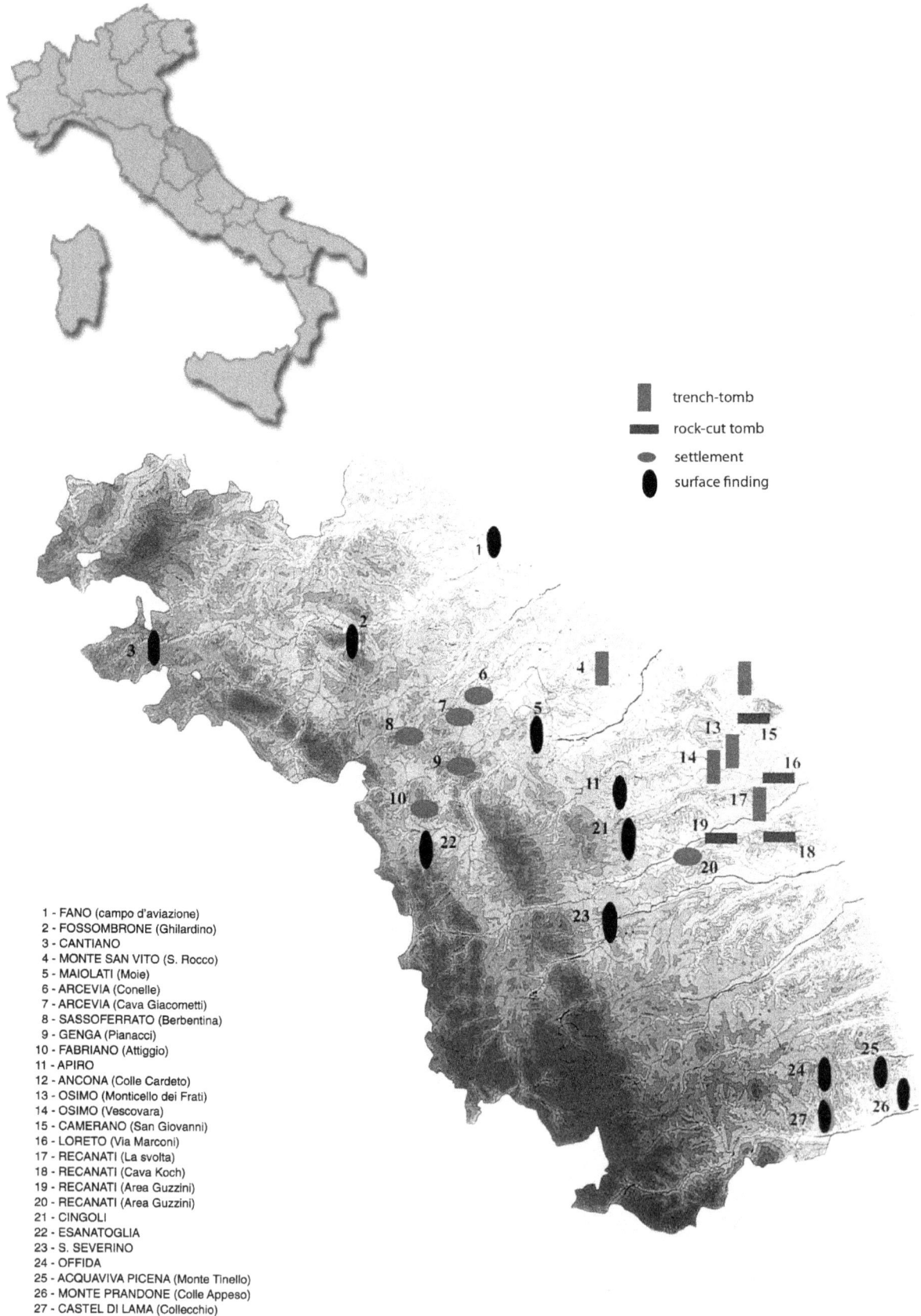

trench-tomb

rock-cut tomb

settlement

surface finding

1 - FANO (campo d'aviazione)
2 - FOSSOMBRONE (Ghilardino)
3 - CANTIANO
4 - MONTE SAN VITO (S. Rocco)
5 - MAIOLATI (Moie)
6 - ARCEVIA (Conelle)
7 - ARCEVIA (Cava Giacometti)
8 - SASSOFERRATO (Berbentina)
9 - GENGA (Pianacci)
10 - FABRIANO (Attiggio)
11 - APIRO
12 - ANCONA (Colle Cardeto)
13 - OSIMO (Monticello dei Frati)
14 - OSIMO (Vescovara)
15 - CAMERANO (San Giovanni)
16 - LORETO (Via Marconi)
17 - RECANATI (La svolta)
18 - RECANATI (Cava Koch)
19 - RECANATI (Area Guzzini)
20 - RECANATI (Area Guzzini)
21 - CINGOLI
22 - ESANATOGLIA
23 - S. SEVERINO
24 - OFFIDA
25 - ACQUAVIVA PICENA (Monte Tinello)
26 - MONTE PRANDONE (Colle Appeso)
27 - CASTEL DI LAMA (Collecchio)

Figure 1 Chalcolitic sites of Marche region; map reworked from the figure at pag.8 of the publication: "Silvestrini M. and D.G. Lollini 2002. Camerano (Ancona) – Località San Giovanni. Necropoli eneolitica. In M. Silvestrini and D.G. Lollini (eds.), Museo Archeologico Nazionale delle Marche. L'Eneolitico, 23-27. Soprintendenza per i Beni Archeologici delle Marche Editore, Falconara".

The archeological context

The burials of Camerano (AN), Osimo - Vescovara (AN), Loreto (AN) and Recanati – Cava Koch and Recanati La Svolta (MC) represent a substantial part of funerary evidences of the Marche region and are all located nearby the coastal area. They belong to the first half of the 4th millennium cal. BC. and, except for the two trench-tombs of Loreto and Recanati – La Svolta, they are all rock-cut tombs (Silvestrini and Lollini 2002; Baroni and Recchia 2006) (Table 1).

Many burials are single (Osimo – Vescovara t.1, t.4, Loreto and Camerano t. 11, t.17, t.46); some contain two individuals (Recanati – La Svolta and Camerano t.33, t.94) or more (Recanati – Cava Koch t.2; Camerano t.21).

The anthropological study of the skeletal remains has highlighted the presence of male individuals of various ages and a reduced presence of female individuals (A.Coppa in Baroni and Recchia 2006) (Table 1).

The methodology

Use-wear analysis was carried out in three steps (*see also* Rots, 2010 pp. 29-35 *for a detailed explanation of the method*). A preliminary macroscopic overview (without the use of microscopes) was carried out to define the degree of preservation of the lithic artefacts by observing the edge-damages and patina, and to classify the most invasive arrowhead fractures caused by an impact.

The identification of the less evident macro- and micro-traces was obtained with a stereomicroscope Nikon ST with 10X oculars, a 1X objective and a reflected light system (for macro-traces) and a metallographic microscope Nikon Eclypse with 10X oculars, 10X and 20X objectives and a reflected light system (for micro-traces).

To begin with, the archeological artefacts were cleaned with absolute ethanol and a tissue and then analyzed with a stereomicroscope in the Archaeological Museum of Ancona where they are stored.

The micro-traces analysis was carried out in the Laboratory of Technological and Functional Analysis of Archeological Artefacts at the University of Rome "La Sapienza" using silicon (Provil Novo Light Fast, Heraeus) moulds on the surface of the lithic tools.

The collection of lithic replicas available in the same laboratory was used as a reference for the interpretation of the archaeological tools.

Use-wear analysis results

Funerary context of Camerano (AN) (Tables 1-2)

All the individuals – in total twelve - found in the six burials that form the funerary context are male – one child, five young adults, four adults and one elderly individual - except for an adult female buried in the multiple tomb n. 21.

Flint arrowheads with barbs and tangs, flint blade knifes, and flint daggers (alone or in combination) are present in the kit of each burial, together with pottery and projectile bone points, the so-called "*punte a taglio sbiecato*".

Only in one case, the senior male of the tomb n. 33, the kit consists of a single arrowhead; all the other individuals

Table 1 Chalcolithic funerary contexts of Marche region considered in this paper;

Site	*Tomb n.°*	*Chronology*	*Context*	*Buried*
Camerano	94	first half 4th millennium cal. BC*	Single rock-cut tomb	1 mature male
Camerano	94	first half 4th millennium cal. BC*	Single rock-cut tomb	1 mature male
Camerano	17	first half 4th millennium cal. BC*	Rock-cut tomb	1 adult male
Camerano	46	first half 4th millennium cal. BC*	Rock-cut tomb	1 sub-adult
Camerano	33	first half 4th millennium cal. BC*	Rock-cut tomb 1th deposition	1 elder male
Camerano	33	first half 4th millennium cal. BC*	Rock-cut tomb 2nd deposition	1 adult male
Camerano	21	second half 4th millennium cal. BC*	Rock-cut tomb	4 young males/ 1 mature female
Camerano	11	second half 4th millennium cal. BC*	Rock-cut tomb	1 young male
Osimo-Vescovara	1	first half 4th millennium cal. BC*	Trench tomb	indeterminable
Osimo-Vescovara	4	first half 4th millennium cal. BC*	Trench tomb	indeterminable
Loreto - 1971		?	Rock-cut tomb	1 individual
Recanati La Svolta 1968		second half 4th millennium cal. BC**	"Trench tomb"	2 adult males
Recanati - Cava Koch	2	first half 4th millennium cal. BC**	Rock-cut tomb	2 individuals

* (Cazzella and Silvestrini 2005);
** (Carboni et al. 2005)

Table 2 Chalcolithic funerary contexts of Marche region considered in this paper; composition of the funerary kit

Site	Tomb n.°	Pottery	"Punta a taglio sbiecato"	Ornaments	"ascia a martello"	Flint dagger	Flint knife	Arrowheads
Camerano	94	X		X				
Camerano	94	X					N.1	N.2
Camerano	17	X	N.1				N.1	N.4
Camerano	46	X	N.5	X				N.3
Camerano	33						N.1	N.1
Camerano	33	X						N.5
Camerano	21	X					N.1	
Camerano	11					N.1		
Osimo-Vescovara	1				N.1	N.1		N.6
Loreto - 1971		X						N.2
Recanati La Svolta 1968		X						N.4
Recanati - Cava Koch	2	X						N.1

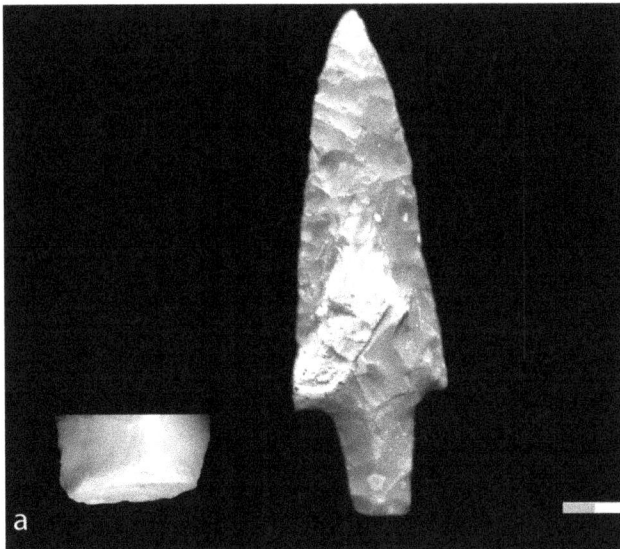

Figure 2 Arrowhead (n° inventory 31343) of the kit of the adult male of tomb 33 from Camerano; a, snap edge-removals observed on the tang interpreted as impact fractures. Scale bar equal to 1cm.

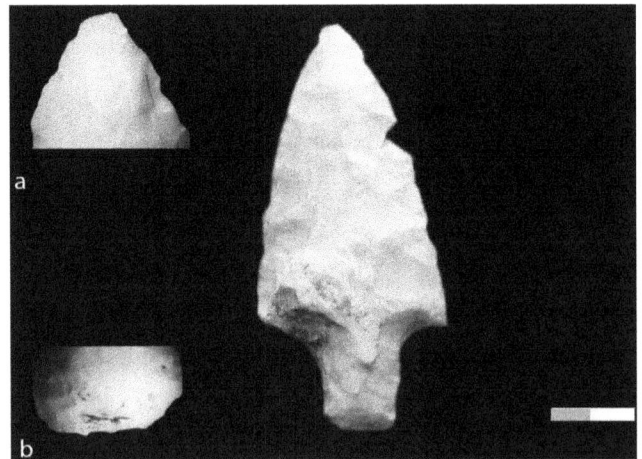

Figure 3 Arrowhead (n° inventory 31345) of the kit of the adult male of tomb 33 from Camerano; a, bending+feather edge-removals observed on the tip and b, snap+ crushing edge-removals observed on the tang interpreted as impact fractures. Scale bar equal to 1 cm.

have a little group of arrowheads, reaching the maximum of five in the same tomb n. 33 (Figures 2,3), in association with an adult male.

Thirteen of the fifteen points collected in this funerary context show small edge-removals testifying frontal impact on tips and especially on tangs. These points were thrown, maybe many times each, without suffering major damage and could have been further used. The functionality of the Camerano burial arrowheads is proven in four other cases by the re-sharpening of tip, tang or barbs (t.46 and t.94).

Macro and micro-traces show a localized abrasion on the medial portion of eleven arrowheads and this endorses the

idea that the arrowheads were fixed to a shaft and likely collected in a quiver, ready for use.

The "quivers" held between two and five arrowheads each, with the exception of the elderly individual to which a single arrowhead was associated. In this last case, the suspicion that the point wounded this individual and was still laying in the buried body must be taken in account. Nevertheless, no concrete data can validate this hypothesis as the position of the arrowheads and the point were lost when the body was removed from its original position and replaced with a new individual. In this case, the arrowhead may have been placed next to the body when it was replaced.

Figure 4 Knife (n° inventory 31346) of the kit of the elder male of tomb 33 from Camerano; a, polishes observed on one edge of the tool and interpreted as hide cutting. Scale bar equal to 1 cm.

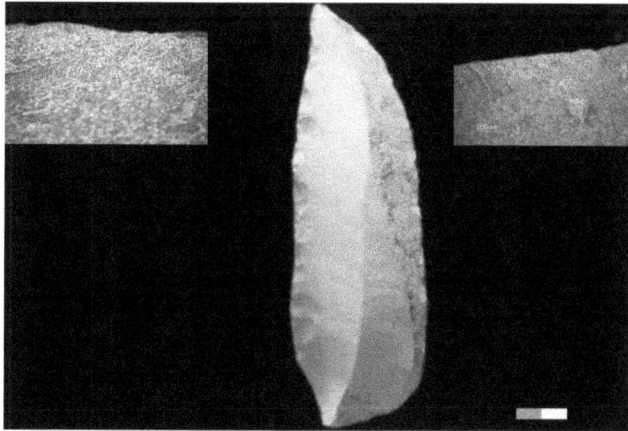

Figure 5 Knife (n° inventory 36537) of the kit of one of the two mature males of tomb 94 from Camerano. Scale bar equal to 1cm; a, polishes observed on one edge of the tool and interpreted as cutting of siliceous herbaceous plants, b, polishes observed on the cortical edge of the tool and interpreted as prehension traces. Scale bar equal to 1 cm.

Regarding the burial of the very young individual (t.46) it is worth mentioning that his kit consisted of five "*punte a taglio sbiecato*" and two arrowheads. Both osseous (Cristiani and Alhaique 2005) and flint arms do not show traces of use. Moreover, the flint arrowheads show traces of re-sharpening. This data may suggest that they were part of the kit of other adult individuals that prepared a proper kit for the "child warrior and hunter" depriving themselves of some of their weapons.

Regarding the flint blade knives, it should be noted that the four items found at Camerano (t.17, t.21, t.33, Figure 4, t.94, Figure 5) show traces of intense usage.

They were hafted and used especially for cutting plants,

both herbaceous and wooden, with the exception of the t.33 knife (Figure 4) that was used to cut mostly hide and meat. Various activities were carried out with these artefacts; these activities left overlapping traces on the edge surface that were partially removed by the re-sharpening of the tools.

Though intensely employed, these knives were still usable when buried. The fact that the close descendants decided to abandon tools that could prove useful in their own kit suggests that the knives were closely related to the identity of their owners.

At this point, the question that can be raised is if the role of the individual was displayed by the possession of a very well shaped tool or, instead, by its function and use.

It is a matter of fact that these blades were the product of skilled work of craftsmen that managed sophisticated technologies based on the indirect percussion or the pressure technique with lever (Guilbeau 2010, unpubl.).

These blades were probably part of an exchange network of goods and they were not accessible to everybody. The owners of such blades benefited from a symbolic distinction for their great effectiveness. In this respect, two knifes from t.17 and t. 33 (Figure 4) show traces that suggest different activities with different materials, as in a kind of "Swiss penknife". Conversely, the knifes from t. 21 and t.94 (Figure 5) show more selective use-wear traces, interpreted as the result of cutting action on siliceous herbaceous plants – possibly reeds – and wooden fibers. In the latter case, the tools were probably used for specialized activities, weaving for example, that may have played a distinctive role in the communities under examination.

Finally, it should be noted that the child and the adolescent of, respectively, tombs 46 and tomb11 do not have knifes in their kit. It appears that these tools may refer only to adults and, in this respect, it is likely that the knife from t.21 belonged to the adult female rather than to one of the young males in the same burials.

These tools were "symbols of agency", displaying the role that their owners played in their communities via their social standing and throughout their whole adult lives.

The only dagger (Figure 6) found in the funerary context of Camerano belongs to an adolescent (t.11); this exceptional tool was the only component of the kit of the buried individual. Use-wear analysis made it clear that the dagger was not used and not hafted. This tool was just preserved for a certain period of time in a bag (*for a detailed description of these traces see* Rots 2010), probably made out of hide, as the localized polishes on tip, barbs and edges testify.

The lack of use-wear suggests that this specific tool had a role only during the funerary rituals. Moreover, its morphology is undeniably similar to that of metal daggers

Figure 6 Dagger (n° inventory 31407) of the kit of the young male of tomb 11 from Camerano; a, polishes observed on the edges of the tool interpreted as the light abrasion produced by the contact with hide, probably a bag where the dagger was maintained. Scale bar equal to 1 cm.

and halberds found in others Chalcolithic contexts and, as the latter tools, may have been a strong symbol of martial display.

The fact that this symbol was associated to an individual that was just entering his adult life as a hunter and warrior suggests that his social role, and that of the group to which he belonged, was very important. The flint dagger buried with the young man was probably an affirmation of the group position within the community.

The Osimo-Vescovara burial (Tables 1-2)

The Osimo-Vescovara funerary context consists of two trench-tombs, only one having a lithic kit. This burial (t.1) is rich in lithic objects: five arrowheads, one polished axe and one dagger.

The arrowheads were partly used and re-sharpened (three items) and partly unused (two items). The latter group could testify that the offering of new arrowheads took place during the funerary ritual. The unused arrowheads enriched an already very conspicuous kit, suggesting that an important individual was buried in this tomb.

The polished axe, of the so-called *"ascia a martello"* type, has traces both of re-sharpening and use after re-sharpening. The tool was in use for a long time and was still perfectly functional at the time of the deposition. The buried individual was honored by his offspring with a tool that was created by means of a very time-consuming technological procedure, and could be still used for many years. Giving up such an object during a funerary ritual symbolizes the prestige of the dead and the connection with his family through a tool that could have been part of the identity of the group for generations.

Another strong prestige symbol was certainly the dagger found in the same burial. As for the Camerano dagger of t.11 (Figure 6), the item from t.1 of Osimo-Vescovara was not hafted and not used. The slight rounding observed on the edges is probably due to the fact that the tool was contained in a bag or a piece of hide or textile. The integrity of the dagger was carefully maintained until its deposition for which this specific tool was probably intended.

The Loreto burial (Tables 1-2)

The funerary kit of the Loreto rock-tomb consists of a retouched and hafted blade that shows traces of intense use. Use-wear traces overlap each other and are interpretable as the working of both plants and animal tissues. Two arrowheads were also part of the same kit; one of them shows traces of impact on its tang.

The Recanati - Cava Koch burial (Tables 1-2)

The Recanati – Cava Koch funerary context consists of seven burials. Only tomb n.2 holds a lithic arrowhead that shows localized and small spots of polish, interpreted as the contact with hide, a bag or a quiver. The arrowhead also has re-sharpening traces on its tang that was successively fractured by impact.

The Recanati – La Svolta burial (Tables 1-2)

In the Recanati-La Svolta trench-tomb two male individuals were found. Besides pottery, a retouched blade made by pressure with a lever and four arrowheads were part of the kit.

The blade knife, as usual, was used for a long period of time for different activities. All the arrowheads have impact traces either on the tip (2 items) or on the tang (3 items).

Discussion

The results obtained through use-wear analysis suggest that the lithic artefacts forming the funerary kit of the Chalcolithic burials of the Marche region had a variety of meanings associated both to the individual and the group to which he belonged.

As highlighted in previous articles (Conati *et al.* 2010; Conti *et al. in press*) arrowheads were part of the identity of each adult male, symbolizing his belonging to the hunters and warriors community.

We can easily imagine that bow and arrows followed their owners from the adolescence to their adult life. As the morphological variability of the arrowheads suggests, there was no craftsmanship in their production. Probably different people knapped these tools, maybe every adult male knapped and/or re-sharpened his own arrows.

Given the limited amount of the arrowheads found in

association to each buried skeleton, it is likely that a part of the content of the quiver was redistributed among the offspring or, more in general, among the male components of the community. Instead, arrows were probably added to the funerary kit of the child (Camerano t.46) and the elderly man (Camerano t.33) since either were respectively not yet and no longer part of the community of hunters and warriors (*see also* Conti *et al. in press*).

For these Chalcolithic groups, lithic arrowheads were a symbol of identity, of "full agency" within the community, that did not have an equivalent in any metal object.

We can assume that, during Chalcolithic times, metal was not so abundant to be used for this kind of weapons that can easily be lost during hunting or battle. This was probably the "practical" reason that made the diffusion of metal arrowheads possible only in later periods.

Therefore, at least during Chalcolithic times, it appears that flint was the favored raw-material for a "key" artefact in terms of social meaning, the arrowhead, that combines both identity and affiliation.

Regarding the flint knifes and the single polished axe found in the funerary contexts of Camerano and Osimo-Vescovara, use-wear analysis testifies that these items played an active role in the daily life of these Chalcolithic groups. They were used for a long time probably passed from one generation to another and, although still effective, they were buried with their owners. These tools were undoubtedly "big gifts" symbolizing the social standing of the dead. In this respect, they had a role similar to that of the still usable metal axes found in other Chalcolithic contexts (Dolfini, 2010). All these types of objects, unlike arrowheads, are based on specialized technologies employed by a restricted number of craftsmen. They were probably part of an exchange network of "goods" and their role was "practical" as working tools as well as "symbolic", since they were "uncommon" artefacts signaling, in some cases, the "signs" of specialized activities, as for example, the knifes of the Camerano tombs t.21 and t.94.

Why did these Chalcolithic groups in the Marche region not put metal knifes or metal axes in the graves of their dead?

Whilst sure evidence is not abundant, metal was known and used in the area (Palmieri and Cazzella 1999). Moreover, the technological analysis of the osseous artefacts found in the Chalcolithic settlement of Conelle di Arcevia (AN), located in a more internal area of the same region, proves that these implements were made using metal tools (Cristiani and Fecchi 2003). This data suggest that metal had a widespread diffusion in these settlements and it was used for mundane activities.

Apparently, the Chalcolithic groups of this region had deep traditional roots in stone technology, facilitated by an easy access to raw material sources that made the transfer of symbolic meanings from stone to metal difficult.

The "resistance" of the stone tradition could explain the presence, in the same burials, of flint daggers whose morphology is extremely similar to the morphology of metal daggers and halberds found in other Chalcolithic contexts.

The martial significance embedded in metal daggers and halberds displayed by other groups that the community under examination directly met, or that they indirectly had heard of, was transferred in flint "copies", a raw material holding a great symbolic power for these communities.

These flint daggers were expressly produced for funerary rituals, just as their metal "cousins", by specialized craftsmen, and were carefully preserved until the occurrence of funerary events related to individuals and groups of particular relevance in the community.

Conclusion

The case study of the Marche region funerary contexts highlights the symbolic value that flint and, more generally, stone acquired from the beginning of Copper Age for these communities. The social value of artefacts such as arrowheads, knifes, axes, and daggers was rooted in a long tradition of stone craftsmanship that, in this geographic area, was facilitated by the proximity to good raw material sources.

The complex social and economic changes that occurred during the late Neolithic (see Conati Barbaro *in this volume*) led these Central Italian communities to "deeply ritualize" stone objects while their day to day activities were more and more characterized by various types of metal objects.

However, this picture is just a "part of the whole story".

"*As a matter of fact, this new material – copper – should not be seen as an alternative to stone or bone/antler: as already pointed out by Rosen (1996), the replacement of the stone tool-kit with the metal one does not follow a simple unilinear process. Both technical systems are linked to different needs, procedures and goals, which are regulated by social and economic factors.*" (Conati Barbaro *in this volume*).

Despite sharing a cultural know-how, the various Chalcolithic communities of the Italian peninsula processed the innovation of metal technology in different ways, that form a variegated picture of social structure that still needs to be investigated in many of its aspects.

Acknowledgments

I am very grateful to Dr. Mara Silvestrini (Archaeological Superintendence of Marche) who gave me the opportunity to study the funerary Chalcolithic collection on exhibit at the Archaeological Museum of Ancona (Marche Region).

References

Baroni, I. and G. Recchia 2006. La necropoli eneolitica di Camerano (Ancona). Atti del settimo incontro di studi di Preistoria e Protostoria dell'Etruria. II, 329-339.

Carboni, G., Conati Barbaro, C., Manfredini, A., Salvadei, L. and M.Silvestrini 2005. La necropoli eneolitica di Fontenoce-Cava Kock (Recanati, Macerata): nuovi dati per l'inquadramento cronologico-culturale, , Atti XXXVIII Riunione Scientifica IIPP, "Preistoria e protostoria delle Marche" vol.II, 949-954.

Cazzella, A and M. Silvestrini 2005. L'Eneolitico delle Marche nel contesto degli sviluppi culturali dell'Italia centrale, Atti XXXVIII Riunione Scientifica IIPP, "Preistoria e protostoria delle Marche" vol.I, 371-386.

Cocchi Genik, D. 1996. *Manuale di preistoria III. L'età del rame*. Octavo Franco Contini Editore, Florence, 325-795.

Conati Barbaro, C., Lemorini, C. and E. Cristiani 2010. The lithic perspective: reading Copper age societies by means of techno-functional approach. *Human Evolution*, 25,1-2, 143-154.

Conti, A., Lemorini, C. and M.Massussi (in publication). La selce si usa, non si "spreca". Uso funzionale e uso rituale dell'industria litica della Necropoli di Selvicciola (Ischia di Castro-VT), Atti XLIII Riunione Scientifica IIPP, "L'Età del Rame in Italia", Bologna 26-29 Novembre 2008.

Cristiani, E. and F. Fecchi 2003. I manufatti in materia dura animale: l'inquadramento tipologico ed i risultati dell'analisi tecno-funzionale. In A. Cazzella, M. Moscoloni and G. Recchia (eds.), *Conelle di Arcevia. Tecnologia e contatti culturali nel Mediterraneo Centrale fra IV e III Millennio A. C. II. I Manufatti in pietra scheggiata e levigata, in materia dura animale, in ceramica non vascolare; il concotto*, 423-502. Edizioni Stampa dell'Ateneo, Roma.

Cristiani, E. and F. Alhaique 2005. Selce o metallo? Approccio sperimentale all'analisi delle modalità di manifattura degli strumenti in materia dura animale presso Conelle di Arcevia (Ancona). Atti XXXVIII Riunione Scientifica IIPP, 939-943.

Dolfini, A. 2011. The function of Chalcolithic metalwork in Italy: an assessment based on use-wear analysis. *Journal of Archaeological Science* 38 (5), 1037-1049.

Guilbeau, D. (unpubl.). Les grandes lames et les lames par pression au levier du Néolithique et de l'Énéolithique en Italie. Thèse de Doctorat. Université Paris Ouest, sous la direction de C. Perlès.

Palmieri, A. and A. Cazzella 1999. Capitolo XI. I manufatti metallici. In A. Cazzella and M. Moscoloni (eds.), *Conelle di Arcevia. Conelle di Arcevia, un insediamento Eneolitico nelle Marche. I. Lo scavo, la ceramica, i manufatti metallici, i resti organici*, 205-208. Gangemi Editore, Roma.

Rots, V. 2010. *Prehension and Hafting Traces on Flint Tools: A Methodology*, Leuven University Press, Leuven.

Silvestrini, M. and D.G. Lollini 2002. Camerano (Ancona) – Località San Giovanni. Necropoli eneolitica. In M. Silvestrini and D.G. Lollini (eds.), *Museo Archeologico Nazionale delle Marche. L'Eneolitico*, 23-27. Soprintendenza per i Beni Archeologici delle Marche Editore, Falconara.

Producing for the Dead, Using while Alive: Lithic Tools Production and Consumption in the Late Neolithic of North-Eastern Iberia

Xavier Terradas

CSIC – IMF. Department of Archaeology.
Egipcíaques, 15. 08001 Barcelona (Spain)

Juan F. Gibaja

Universidade do Algarve – Faculdade de Ciências Humanas e Sociais.
Campus de Gambelas. 8000-117 Faro (Portugal)

Antoni Palomo

Universitat Autònoma de Barcelona – Arqueolític.
Sant Martirià, 56. 17820 Banyoles (Spain)

Corresponding author: terradas@imf.csic.es

Looking backwards

In the north-east of the Iberian Peninsula, the lithic assemblages of post-Neolithic periods have been dealt with so superficially that the most note-worthy features that have attracted archaeologists' attention, and which have therefore been reflected in publications, only refer to typological and morphometric aspects. In this respect, when referring to the known lithic record, especially the objects found in funerary contexts, mention is only made of the raw material used, its colour, size and its morphological determination. This determination is based on the comparison with contemporary implements or on parallels with morphotypes used by archaeologists working in other disciplines. Thus, it is customary to find citations of blades, knives, halberds, daggers, etc. In conclusion, it was the pieces that were considered exceptional from the stylistic point of view that were sought out, studied and published, whereas other kinds of products that did not follow these stylistic canons were discarded or ignored, although they could in fact be more illustrative of the subsistence behaviour of the societies being studied.

However, there have been other researchers with different proposals, whose objectives went beyond the mere description of certain objects. One of the most outstanding examples is the research carried out by Salvador Vilaseca in the Tarragona area for many years. He studied multiple archaeological sites, of both primary and secondary character (the famous surface lithic assemblages he considered as flint-knapping workshops). In his publications, S. Vilaseca (1973) did not only consider aspects related with the morphology of the pieces, but also

used his knowledge of the geology of the area to establish the possible areas of provenance of the raw material used. Although his proposals about the origin of the raw materials were evidently made *a visu*, they have become a highly valuable source of information for later research.

In this context, when we began our own research in 2004, we encountered a precedent of extremely limited studies, where we practically had to start again from scratch. This surprised us, as the implements found with burials in the north-east Iberian Peninsula in the final Neolithic-Chalcolithic are not only of exceptional quality, but also include morphotypes some of which are unique and had never been studied before. We are referring to the long flint blades and the various arrowheadsthat are often found in association with them.

We had to make a new beginning: determine the provenance of the raw materials with which the stone tools were manufactured, carry out more rigorous morphotechnical studies whose technological aspects should help us to understand the complex technical systems used in the production of these blades and points, and undertake functional analysis in order to know whether these tools deposited in tombs had been used or not, and what that use may have been. Although we have now been working on this topic for several years, despite the progress made (Clop Gibaja *et al.* 2001; 2006; Gibaja Palomo *et al.* 2004; 2005; Palomo Terradas *et al.* 2004; Terradas Palomo *et al.* 2005), many problems remain to be solved.

At the same time, if we aim to know what the lithic tool-kit used by communities of that period was, we should also

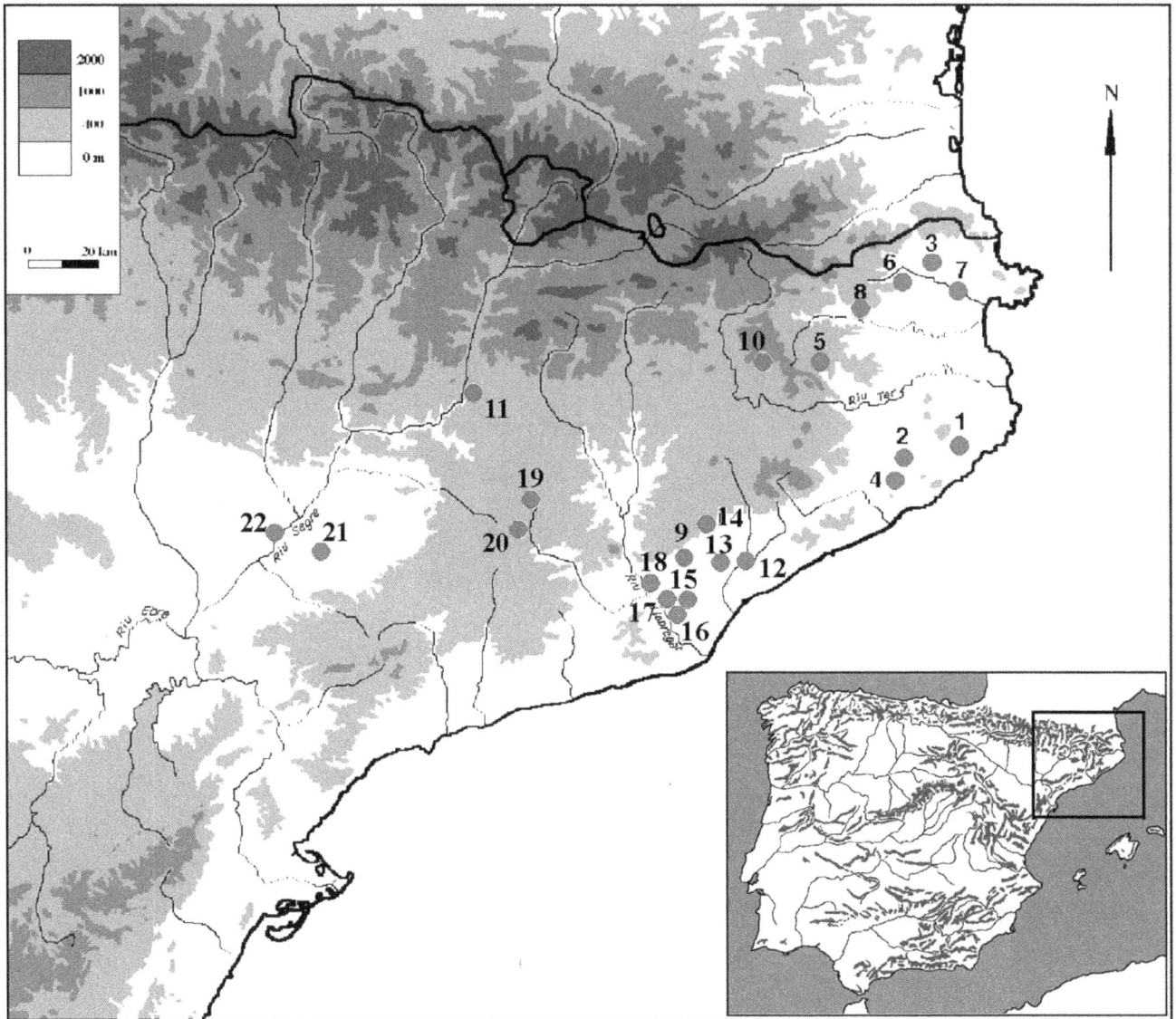

Figure 1: Location of the sites mentioned in the text in the context of the Northeast of the Iberian Peninsula: 1. Mas Bousarenys, 2. Llobinar, 3. Cabana Arqueta, 4. Dolmen de Pericot, 5. Cementiri dels Moros, 6. Fontanilles, 7. Vinya del Rei, 8. Encantades de Martís, 9. Vapor Gorina, 10. Bauma del Serrat del Pont, 11. Collet del Sàlzer, 12. Costa de Can Martorell, 13. El Coll, 14. Can Filuà, 15. Bòbila Madurell, 16. Calle Paris de Cerdanyola del Vallès, 17. Sitges UAB, 18. Can Roqueta, 19. El Collet de Brics d'Ardèvol, 20. Dolmen de Les Maioles, 21. Minferri et 22. Les Roques del Sarró.

study domestic contexts; i.e. the artefacts usually found in silos and waste pits, which are common structures at many sites in north-east Iberia. Otherwise, if the materials discovered in tombs were specially selected tools to be used as grave goods, as a kind of offering, the overall picture of lithic tools in the period would be totally distorted. For that reason, in the present paper we will make constant references to the lithic implements recorded at some of the few settlements that are known in the area of study.

Chronocultural Background

Between the end of the 4th Millennium and the first centuries of the 2nd Millennium cal BC, a series of gradual but deep transformations took place among the communities in the northeast of the Iberian Peninsula. These affected their social and economic organisation and

also their ideological concepts and customs. These changes were based on the progressive consolidation of a farming economy which led to the development of strategies to organise and control production, within the context of an increasingly sedentary population. This in turn led to an intensification in agriculture (cultivation of dry land, use of the plough) and pastoralism (exploitation of secondary products: milk, wool, animal traction).

However, this intensification in animal husbandry was not oriented towards any one particular species, and instead all the main domestic species were kept in a diversified and complementary manner. At the same time, wild animals were hunted as a secondary dietary contribution.

The settlements were small and disperse, located in the lowlands or in areas of gentle relief, always near rivers or other water bodies (Figure 1). Habitation structures have been identified at some of them. They have a rectangular or sub-circular floor plan, frequently excavated in the ground, and with post-holes to support two-sided roofs. These houses were used for only a short time, and are found in isolation or forming small groups. In no case have any structures been found that might suggest a possible defensive concern. Equally, it has been shown that mountain areas were re-occupied, sometimes in areas at over 1000m above sea level. In these places, caves were used as temporary shelters or as storage spaces.

Therefore, they may have been small groups, characterised by their great mobility, who occupied different settlements. Their subsistence was based on farming and fishing with significantly diversified domestic species, which were exploited in a balanced way, making use of the most favourable conditions in each ecological niche. However, the specific functionality of each site and their relationships with each other remains to be determined.

These strategies can be detected at the end of the middle Neolithic, and resulted in a progressive division and social specialisation of labour, as well as the appearance of incipient power structures and violent inter-group conflicts.

However, this process does seem to have been too intense in the northeast of the Peninsula, at least in comparison with neighbouring areas (SE France, Valencia and, above all, the southeast of the Peninsula). The explanation of this difference could lie in the absence of mineral resources that are suitable for metallurgical production in the area. This fact is reflected in the archaeological record, where metallic artefacts are rare and are normally associated with funerary contexts. Owing to their rarity, and the scarcity of the raw materials used, these implements have been interpreted as symbols of the social power held by certain individuals.

Yet the particular situation in the northeast of the Peninsula, away from the main routes along which metal artefacts were in circulation, did not mean the area was isolated; on the contrary, a series of materials have been found which, because of their singularity and geographical distribution, are representative of new, large inter-group contact networks. Among these materials, some of the most important are precisely long blades and arrowheads made in flint (Clop Gibaja *et al.* 2001; 2006; Palomo and Gibaja 2002; Gibaja Palomo *et al.* 2004; Palomo Terradas *et al.* 2004; Terradas Palomo *et al.* 2005). These implements usually appear forming part of grave goods in funerary contexts. However, these tools bear little comparison with those found at settlements. At the latter kind of site, the artefacts found are extremely expeditious, characterised by flakes knapped from raw materials of local origin.

On the other hand, the fact that we occasionally have to work with archaeological assemblages that come from old excavations means that we have no precise chronological reference for them. And so it is, in the case of many megalithic burials or burial caves that we have studied, the chronological attributions have been based, not on absolute radiocarbon determinations, but on archaeological parallels usually in relation with the contents and containers of the tombs. Fortunately, this situation has changed in recent years, as is shown by the series of absolute dates obtained in the latest archaeological fieldwork, especially at settlements (Table 1).

Implements found at burial sites: Long Blades and Arrow-heads

The beautiful long flint blades and the outstanding arrowheadsthat have always attracted researchers since the 1950s (Pericot 1950), are in fact the most representative implements in funerary contexts corresponding to the period: megalithic burials, burial caves and hypogea. However, two important questions must be highlighted: the first is the repeated presence of small flint flakes, fragments and even knapping remains at certain burials, theoretically in the same levels of the deposit. The second, as already mentioned above, is that the characterisation of the lithic assemblage of the period cannot be made, ignoring the evidence found at settlements.

In the case of burials, the available archaeological information for each site is extremely varied, as some were excavated in the middle of the last century, whereas others have been studied recently. To sum up, we have examined (Figure 1):

1. A total of 30 whole or fragmented blades from the sites of Mas Bousarenys (Santa Cristina d'Aro, Girona), Llobinar (Fitor-Fonteta, Girona), Dolmen de Pericot (Torroella de Montgrí, Girona), Cabana Arqueta (Espolla, Girona), Cementiri dels Moros (Torrent, Girona), Vinya del Rei (Vilajuïga, Girona), Fontanilles (Sant Climent Sescebes, Girona) and Les Encantades de Martís (Esponellà, Girona).
2. From the artificial cave at Costa de Can Martorell (Dosrius, Barcelona), the collective burial in París Street (Cerdanyola, Barcelona) and the megalithic structure at Collet del Sàltzer (Odèn, Lleida), we have studied a large group of arrow-heads. The case of Costa de Can Martorell is especially important as the site has yielded 68 examples.

Regarding the long blades, preliminary macroscopic analysis has enabled us to distinguish, on the one hand, blades made from chalcedony and other siliceous rocks with megacrystalline granular textures from northeast Iberia (Catalonia), and, on the other, blades knapped from flint with a basically micro or cryptocrystalline granular texture, from outside the area. Although for the moment, we have been unable to identify the exact provenance of the latter materials, we are considering two hypotheses: the French regions of Roussillon-Languedoc-Provence (Briois 1997; Grégoire 2000; Plisson Bressy *et al.* 2006; Renault 1998;

Table 1. Radiocarbon determinations at final Neolithic-Chalcolithic sites in the northeast of the Iberian Peninsula. From Castany Alsina and Guerrero 1992; González Martín and Mora 1999; Martín and Mestres 2002; Mestres 2002.

Archaeological Site	Laboratory Code	Date
El Coll	MC-1242	4775 ± 80 BP
El Coll	MC-2143	4640 ± 90 BP
Bòbila Madurell	MC-1243	3750 ± 90 BP
Bòbila Madurell	UBAR-83	3620 ± 80 BP
Bòbila Madurell	UBAR-399	4020 ± 130 BP
Bòbila Madurell	UBAR-400	3870 ± 110 BP
Bòbila Madurell	UBAR-398	3850 ± 100 BP
Bòbila Madurell	UBAR-87	3350 ± 90 BP
Bòbila Madurell	UBAR-273	3310 ± 60 BP
Bòbila Madurell	UBAR-275	3150 ± 50 BP
Bòbila Madurell	UBAR-277	3140 ± 50 BP
Bòbila Madurell	UBAR-278	3100 ± 60 BP
Bòbila Madurell	UBAR-274	3060 ± 50 BP
Bòbila Madurell	UBAR-286	4030 ± 290 BP
Cova 120 (nivel II)	GIF-6925	4240 ± 70 BP
Cova 120 (nivel I)	UGRA-107	3190 ± 130 BP
Costa de Can Martorell	LY-7837	3810 ± 55 BP
Costa de Can Martorell	LY-7838	3795 ± 55 BP
Bauma del Serrat del Pont (nivel III.1)	BETA-64939	4020 ± 100 BP
Bauma del Serrat del Pont (nivel III.1-III.2)	BETA-79222	4100 ± 70 BP
Bauma del Serrat del Pont (nivel III.2)	BETA-90620	4490 ± 70 BP
Bauma del Serrat del Pont (nivel III.2)	BETA-90621	4460 ± 70 BP
Bauma del Serrat del Pont (nivel III.2)	BETA-90623	4440 ± 50 BP
Bauma del Serrat del Pont (nivel III.3)	BETA-164087	4340 ± 70 BP
Bauma del Serrat del Pont (nivel III.3)	BETA-168715	4430 ± 40 BP
Bauma del Serrat del Pont (nivel II.3)	BETA-69597	3840 ± 90 BP
Bauma del Serrat del Pont (nivel II.4)	BETA-64940	4100 ± 100 BP
Bauma del Serrat del Pont (nivel II.5)	BETA-90622	4200 ± 70 BP
Roques del Sarró	BETA-92207	4670 ± 70 BP
Roques del Sarró	BETA-92206	4040 ± 60 BP
Roques del Sarró	BETA-92205	3950 ± 90 BP
Roques del Sarró	BETA-92208	4830 ± 40 BP
Collet de Brics d'Ardèvol	UBAR-89	3960 ± 60 BP
Minferri	UBAR-548	3590 ± 110 BP
Minferri	UBAR-547	3560 ± 70 BP
Minferri	UBAR-549	3510 ± 60 BP
Minferri	UBAR-550	3450 ± 150 BP
Minferri	BETA-92279	3380 ± 70 BP
Sitges UAB	UBAR-580	3580 ± 60 BP
Sitges UAB	UBAR-579	3520 ± 60 BP
Les Maioles	UBAR-558	3475 ± 50 BP
Les Maioles	UBAR-559	3465 ± 50 BP
Les Maioles	UBAR-560	3495 ± 50 BP
Calle Paris de Cerdanyola		4110 ± 60 BP
Can Roqueta/Diasa	UBAR-230	3370 ± 80 BP
Can Roqueta/Diasa	BETA-91583	3570 ± 140 BP
Can Roqueta /Diasa	BETA-91849	3900 ± 120 BP
Can Filuà	UBAR 555	3500± 50 BP
Can Filuà	UBAR 556	3500± 50 BP

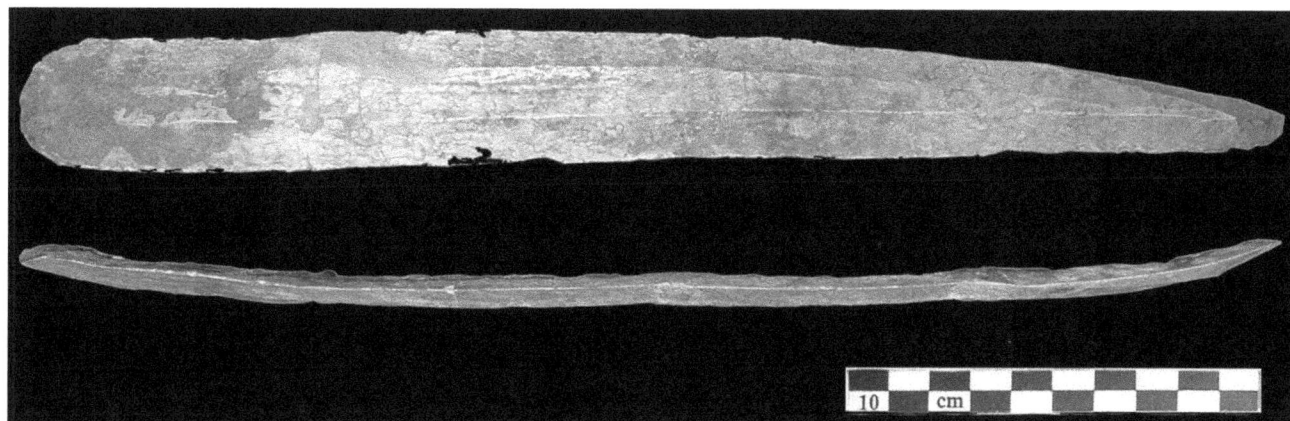

Figure 2. Long blade from Serra de l'Arca, deposited at the Museu Episcopal de Vic.

2006), and the upper-middle Ebro Valley (Ortí Rosell *et al.* 1997).

In order to manufacture these blades, two different techniques were used. Whereas most of them were knapped by indirect percussion, others were made by means of a pressure flaking lever device. Regarding the latter technique, certain morphotechnical characteristics can be identified in some blades, such as an acute dihedral butt, which suggests that copper tips were used to knap them. This question has been analysed fully for archaeological sites in Andalucia, where the research of A. Morgado and J. Pelegrin has been able to test this hypothesis (Pelegrin and Morgado 2007). Although both techniques involve the careful preparation of the core, the result is the production of longer and more robust blades, some of which are over 30 cm long. The clearest example is the blade from the site of Serra de l'Arca (Aiguafreda, Barcelona), deposited in Vic Museum and which we hope to study in depth in the near future (Figure 2).

Normally, the edges of these blades were retouched systematically by a simple, deep, direct and continuous retouch along the lateral edges. After being resharped several times, they progressively acquire an abrupt appearance and a denticulate delineation. Equally, in some cases, the flat invasive retouch has been much more careful, giving the blade a pointed shape, similar to that of a dagger. The most representative example of these blades we have studied is the one also knapped by lever pressure, recovered at the dolmen of Cabana Arqueta.

In both the cases of unretouched blades and those with modified edges, it is clear that blades were not knapped at the settlements that have been excavated so far. We can make this claim based on the absence of any evidence of knapping remains representatives of their manufacture at this kind of site. We therefore tend to believe these products were put in circulation as manufactured blades and that their production probably took place in areas near the sources of the raw materials.

One of the most significant questions which we need to discuss is about the function of these long blades. These beautiful objects have often been attributed a purely symbolic use. This would evidently imply that these blades were in fact unused. We therefore decided to carry out use-wear tests.

On the contrary to what had been thought, most blades had been used. Although cutting and processing cereals was the most common use (Figure 3), we could also identify other activities such as hide working, butchering animals and the transformation of indeterminate minerals. Not only this, some of the blades had been used for the work of different activities. This is the case of certain blades from Mas Bousarenys, Llobinar, Dolmen de Pericot or Cabana Arqueta. This showed us that they were implements with different uses as occurs at other European sites (Plisson Mallet *et al.* 2002).

The use seen on some of the blades used for reaping was so intense that their edges became quite worn and rounded. Occasionally, in order to avoid this rounding, the objects underwent a continuous process by which they were repaired and their lives were prolonged. This process has sometimes been so important that the transversal cross-section of the blade has been completely modified, with the result that much of the original edge has been lost. An extreme case can be seen in the blade from Mas Bousarenys.

The other representative artefacts in funerary contexts are arrow-heads. In the present study we include the materials from the sites of Costa de Can Martorell, París Street and the megalithic monument of Collet del Sàltzer. As these are recent excavations, we think it would be appropriate to make a short introduction to each site. Thus, Costa de Can Martorell is an artificial hypogeum whose funerary chamber takes the form of a semi-circular area of 7m2, excavated in the bedrock, and which is reached along a corridor preceded by a megalithic entrance or ante-chamber formed by lines of flagstones standing vertically on both its sides. In the interior the remains of between 195 and 205 individuals

Figure 3. Blade found at the site of Les Encantades de Martis, with intense marks produced by cereal reaping (photo at 100x)

have been identified. A lithic assemblage consisting, almost exclusively, of 68 arrowheadswas recovered together with the remains (Palomo and Gibaja 2002).

The site in París Street also appears to be a hypogeum excavated in the ground, with an oval floor area, and in which as many as 36 successive burials took place. Most of these burials were found in one of the levels -UE12-, together with eight arrowheadsand abundant ornament elements made from *dentalia* molluscs. A later level -UE5-, contained several burials associated in this case with two bell-beakers pots (Gibaja Palomo *et al.* 2006).

Finally, from Collet del Sàltzer we have studied two flint arrow-heads. This is a megalithic structure with a rectangular chamber formed by three large orthostat flagstones, where a few teeth were found, belonging to three young children (one about 1-2 years old, one 2-3 years old and a third 6-7 years old, approximately) (Castany Baulenas *et al. in press*).

The morphotechnical analysis of this ensemble of arrowheadsfound in different funerary contexts, shows us that it is not a homogeneous ensemble. They were mostly manufactured from flakes, but their shaping by retouch

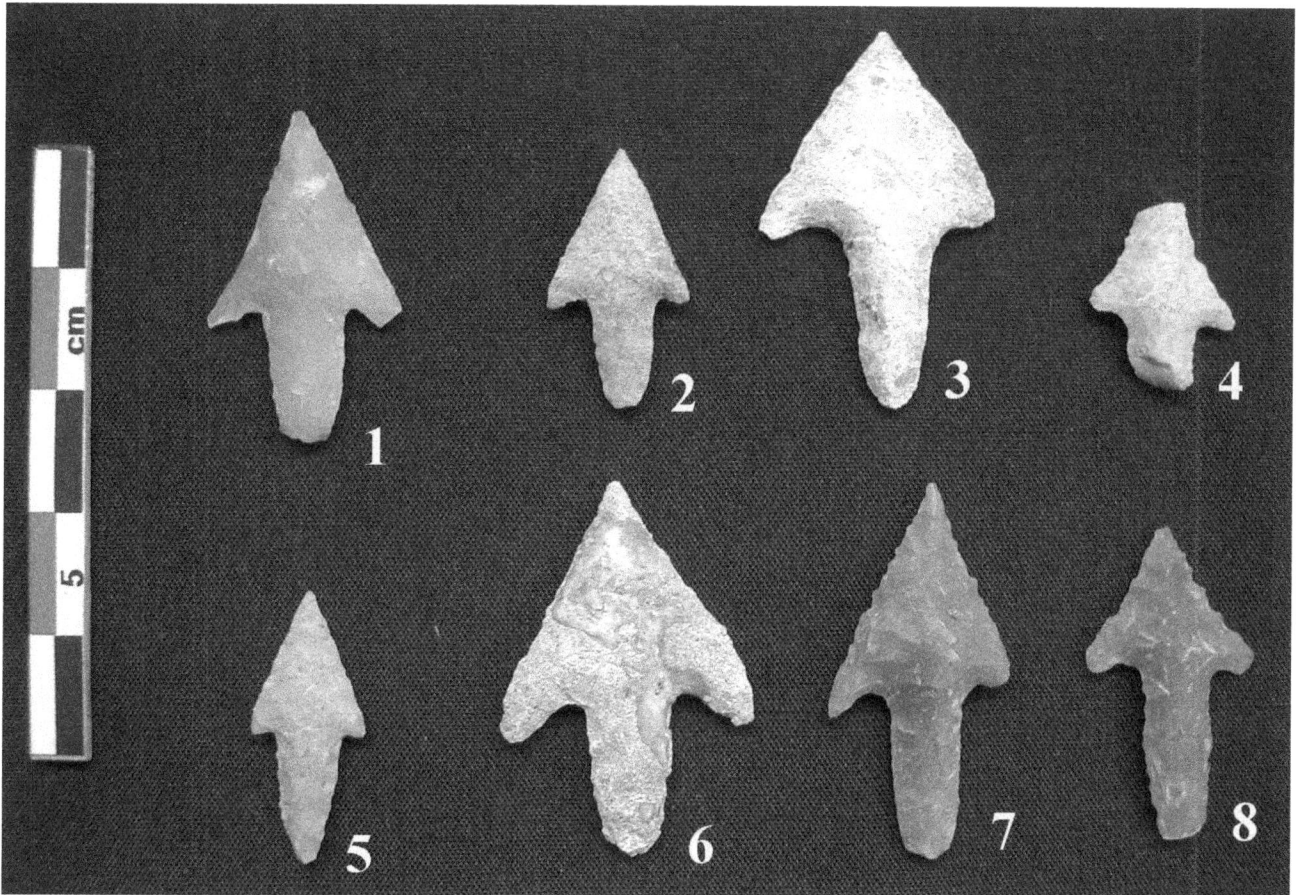

Figure 4. Arrow-heads found in the collective burial at París Street (Cerdanyola, Barcelona)

has enabled us to define three different levels of technical difficulty, probably relating to the skills of the knappers: 1) *Arrowheads of low technical difficulty*, where the scarce retouching that took place was aimed only at regularize their perimeter and not their thickness; 2) *Arrowheads of medium technical difficulty*, where invasive retouching by pressure was able to shape the surface of the artifact symmetrically; and 3) *Arrowheads of high technical difficulty*, shaped by careful laminar retouching made by pressure.

In this respect, we may add that two types of arrow-head are found at Costa de Can Martorell and in París de Cerdanyola Street; the first of these have very long wings and short tang, and are highly efficient when they lodge in the body, and the others have short wings and a long tang, which allow a more accurate and rapid shot, but which, however, easily fall out of the body (Figure 4).

In contrast, at the megalith of Collet del Sàltzer two magnificent foliate points were found, displaying retouch by pressure. This technology is an indication of the artesan's profound technical skills and the great investment in labour in their manufacture.

In any case, it seems significant that nearly 80% of the arrowheads from Costa de Can Martorell are fractured in one or more places. And in addition, out of this 80%, nearly

25% were completely unusable, and therefore impossible to repair, because of the major fractures suffered in the areas of the apex, the wings and/or the tang.

In París Street, however, only one arrow-head displays possible impact fractures in the apical area and the tang. Although in the other specimens no macroscopic impact fractures can be seen, the poor state of conservation of their surfaces has made it impossible to detect other kinds of diagnostic marks, such as impact striations. Therefore we have reached the conclusion that we do not posses sufficient criteria to determine whether they were used or not.

Lastly, the two foliate points from Collet del Sàltzer display probable impact fractures in the apical and/or proximal areas. Nevertheless, to judge from the good preservation of these artefacts, it appears that the two arrowheads were specifically chosen to be left as grave-goods.

All this information leads us to think that, although used arrowheads could be left in the burial as grave-goods, we do not rule out the possibility that on occasions they reached the site lodged in the bodies of the individuals buried there, as a consequence of violent acts. In fact, this hypothesis has been proposed for Can Martorell burials (Mercadal 2003). This circumstance is not unknown at other funerary contexts in northern Iberia, where not only have many

Figure 5. Cores knapped with local raw materials recovered in the Vapor Gorina site (Sabadell, Barcelona)

arrowheadsbeen recovered with impact fractures or broken through their barbs and/or tang, but also individuals with arrowheadslodged in their bones. The most representative examples have been found at Longar -Viana, Navarra- (Armendáriz and Irigaray 1995), Aizibita -Cirauqui, Navarra- (Beguiristain 1996) and San Juan Ante Portam Latinam -La Guardia, Álava- (Vegas 1999; Márquez 2004).

Consequently, the traditional concept that all the materials found in a tomb belonged to the grave-goods deposited with the body is being brought into question, because of these new discoveries. The presence of fractured arrowheadsin burials of this period must contribute to a profound theoretical re-consideration of the concept of grave-goods and of intra- and inter-group social relationships.

Implements found at Domestic sites

The lithic assemblages found in domestic contexts are completely different. For example, in relation with the raw materials, the rocks found near the settlements are used exhaustively. This can be seen in the analysis of the artefacts recovered at sites like El Coll (Llinars del Vallès, Barcelona), Bòbila Madurell (Sant Quirze del Vallès, Barcelona), El Collet de Brics d'Ardèvol (Pinós, Lleida), Bauma del Serrat del Pont (Tortellà, Girona), Can Roqueta (Sabadell, Barcelona), Vapor Gorina (Sabadell, Barcelona) and Can Filuà (Santa Perpètua de Mogoda, Barcelona) (Martín 1979; Castany Alsina and Guerrero 1992; Miret 1993; Díaz Bordas *et al.* 1995; Miret and Martín 1998;

González Martín and Mora 1999; Terradas and Borrell 2002; Roig Molina *et al. in press*), (Figure 1). Objects manufactured from clearly allochthonous material are found very rarely, as, for example at the habitats of Can Roqueta and Can Filuà. At the former a sickle blade made from a flint tablet was found (Gibaja and Palomo 2004) and at the latter, the medial fragment of a long blade was recovered in a silo; this was made from a clearly veined flint, like the kind occurring at Roussillon-Languedoc-Provence or in the provinces in the upper-middle Ebro Valley. But generally speaking, the rocks used at this type of settlement come from the surrounding areas. Thus, both quartz and flint are found in the palaeo-channels that eroded the Miocenic levels at the site of Can Roqueta (Palomo and Rodríguez 2003).

Regarding the technical processes used to produce the tool blanks, these are closely related to the lithological characteristics of the raw material. Therefore, for local rocks with poor conchoidal fracturing, such as thick-grained flint, filonian quartz, quartzite and limestone, direct percussion was used to extract flakes (Figure 5). At sites like Vapor Gorina, Bauma del Serrat del Pont, Bòbila Madurell, El Coll, El Collet de Brics d'Ardèvol or Can Roqueta, this knapping process was carried out almost wholly at the settlements, as is shown by the presence of all the products and waste linked with the different manufacturing stages (Castany Alsina and Guerrero 1992; Díaz Bordas *et al.* 1995; Miret and Martín 1998; González Martín and Mora

1999; Terradas and Borrell 2002; Gibaja 2003; Roig Molina *et al. in press*).

It's important to point out that after the second quarter of the second millennium calBC there is a decline in the number of lithic remains recovered at the sites, accompanied by a near disappearance of retouched blanks. However, and although in general the number of retouched artefacts is small, certain lithologies and implements are retouched more than others. This is the case of blades produced from allochthonous flint, with which points and sickle elements are manufactured especially. In contrast, regarding flakes made from local rocks, these are only occasionally transformed into denticulates, notches or scrapers.

Considering the function of these tools, the few use-wear studies that have been carried out are insufficient to give us an accurate overall view of the characteristics of the artefacts in relation with the activities they were used for. However, two large groups can be distinguished (Figure 1):

1. On one hand, at sites like Bauma del Serrat del Pont, El Coll, Les Roques del Sarró or Vapor Gorina (Alonso Clemente *et al.* 2000; Gibaja 2002; 2003; Roig Molina *et al. in press*), tools have been recovered that were used for such activities as hunting, reaping cereals, butchering animals, hide treatment or wood processing. For these tasks we have been able to define both the production of expedient tools made with flakes derived from local rocks, which have been used for single tasks that take only a short time, and also a much more effective and versatile tool-kit, obtained from flakes and blades made from allochthonous rock. These have been used for different activities or employed in the manufacture of certain tools, such as the points later used as projectiles.
2. On the other hand, at settlements like Minferri, sickle elements are particularly important (Alonso 1999). Most of the implements found at these sites are flakes and blades, occasionally with denticulate edges, that have been used to reap cereals. Judging from the well-developed wear-traces, these are often artefacts with a long life, in use during a long period of time.

Conclusions

Between 3500 and 1500 calBC profound socio-economic transformations took place in northeast Iberia, and these are reflected in the archaeological record.

The lithic tool-kit used by human communities at that time is very diverse, if we take into account the dichotomy between the implements found in funerary contexts and those at domestic sites. Thus, whereas at burials we encounter grave-goods consisting of arrowheadsand long blades made with high quality flint, manufactured by specialist artisans, at settlements the expedient tools are usually made from rock of local origin and used for a wide range of functions.

Precisely because the tools deposited in tombs as grave-goods had been used and re-used practically until they were worn out, we are inclined to believe that apart from their possible symbolic meaning, these long blades had previously been used in different subsistence activities, as well as with other technical activities linked with the manufacture of other artefacts and objects. It is therefore clear that these objects were not of an exclusively ideological character, although at a certain moment they acquired an important symbolic value as funeral offerings.

Tools that were difficult to manufacture required raw materials of high quality, whose provenance lay outside the region, as well as demanding technical skills to produce the best and most effective results. This makes us reflect on the probably restricted control that certain individuals might have held over the production and exchange networks of these raw materials of remote origin. If this hypothesis can be proven, it will be necessary to consider the presence of social control mechanisms over the means of production of the long blades, as well as to assess the forms of access to the possession and use of these goods by part of the population.

Equally, it will be important to continue studying, in greater depth, the lithic assemblages found at domestic sites, as they will provide vital data about the exploitation strategies of mineral resources that were used by these human communities of the final Neolithic-Chalcolithic. It is surprising how much the specialized and curated tools found in tombs contrast with the expeditious kinds of artefacts recovered at habitats. These implements, from settlements, even seem to reflect a loss in the technical tradition that had been forged from generation to generation.

As a result, we may say that after the second quarter of the second millennium cal BC, a turning point occurs in the representation of knapped lithic tools, which could be related, among other circumstances, to the greater importance that metal tools began to acquire. In fact, at Can Roqueta the explanation that has been given for the low frequency of lithic tools is the wide knowledge the groups possessed about bronze metallurgy (Rovira 2003).

Whatever is the truth, much work remains to be done. We are in the initial stage of research. Knowing in depth the areas of provenance of the raw materials used for the production of long blades; their knapping process and ways in which they were put into circulation, as well as why they became common in many other contemporary archaeological contexts throughout Europe (Vaquer and Briois 2006), will help us to gain a more precise and overall vision of how exchange networks were established as well as their social significance.

Acknowledgements

We would like to express our most sincere gratitude to all the institutions and museums that have opened their doors to us so we could study the materials presented in this paper: Museu Arqueològic de Barcelona, Museu Arqueològic Comarcal de Banyoles, Museu Episcopal de Vic, Museu de Sant Feliu de Guixols and Can Quintana Centre Cultural de la Mediterrànea (Torroella de Montgrí). We also like to thank Peter Smith from the translation to english of this paper.

References

Alonso, N. 1999. *De la llavor a la farina. Els processos agrícoles protohistòrics a la Catalunya Occidental*. Monographies d´Archéologie Méditerranéenne 4, Lattes.

Alonso, N., Clemente, I., Ferrer, C., Gené, M., Gibaja, J.F., Juan-Muns N., Junyent, E., Lafuente, A., López, J.B., Llussà, A., Mirada, J., Miró, J.M., Morán, M., Roca, J., Ros, M.T., Rovira, C. and Tartera, E. 2000. Les Roques del Sarró (Lleida, Segrià): Evolució de l'assentament entre el 3600 cal a.n.e. i el 175 a.n.e.. *Revista d'Arqueologia de Ponent* 10, 103-173.

Armendariz, A. and Irigaray, S. 1995. Violencia y muerte en la prehistoria. El hipogeo de Longar. *Revista de Arqueología* 168, 16-29.

Beguiristain, M.A. (1996) Belicosidad en la población usuaria de los dólmenes navarros. Reflexiones y perspectivas. *II Congreso de Arqueología Peninsular*, 323-332.

Briois, F. 1997. *Les industries lithiques en Languedoc méditerranéen (6000-2000 av. JC). Rythmes et évolution dans la fabrication des outillages de pierre taillée néolithiques entre mer et continent*. Thèse de doctorat - EHESS, Toulouse.

Castany, J., Alsina, F. and Guerrero, Ll. 1992. El Collet de Brics d'Ardèvol. Un hàbitat del Calcolític a l'aire lliure (Pinós, Solsonès). *Memòries d'Intervencions Arqueològiques a Catalunya* 2, Generalitat de Catalunya, Barcelona.

Castany, J., Baulenas, A., Gibaja, J.F. and Palomo, A, in press. El megàlit del Collet del Sàltzer I (Odèn, Solsonès). *Actes del 1er Col.loqui d'Odèn. La prehistòria avui en el Prepirineu Lleidatà*.

Clop, X.; Gibaja, J. F.; Palomo, A. and Terradas, X. 2001. Un utillaje lítico especializado: las "grandes láminas" de sílex del Noreste de la Península ibérica. *XXVII Congreso Nacional de Arqueología (Huesca 2003). Instituto de Estudios Altoaragoneses. Bolskan* 18, 311-322.

Clop, X., Gibaja, J.F., Palomo, A. and Terradas, X. 2006. Approvisionnement, production et utilisation des grandes lames en silex dans le nord-est de la Péninsule Ibérique. In J.Vaquer and F. Briois (eds), *La fin de l'Âge de Pierre en Europe du Sud*. Archives d'Écologie Préhistorique 2, 33-246, Toulouse.

Díaz, J., Bordas, A., Pou and R., Martí, M. 1995. Dos estructuras de habitación del Neolítico Final en el yacimiento de la "Bòbila Madurell" (Sant Quirze del Vallès, Barcelona). *1º Congresso de Arqueologia Peninsular*, Trabalhos de Antropologia e Etnologia 35 (1), 17-34, Porto.

Gibaja, J.F. 2002. Anàlisi funcional de les restes lítiques tallades. In G. Alcalde, M. Molist and M. Saña (eds.), *Procés d'ocupació de la Bauma del Serrat del Pont (La Garrotxa) entre 5480 i 2900 cal AC*. Publicacions Eventuals d'Arqueologia de la Garrotxa 7, 81-82, Olot.

Gibaja, J.F. 2003. *Comunidades Neolíticas del Noreste de la Península ibérica. Una aproximación socio-económica a partir del estudio de la función de los útiles lítico*. British Archaeological Reports, International Series 1140, Oxford, BAR Publishing.

Gibaja, J.F. and Palomo, A. 2004. Las hoces líticas usadas durante la prehistoria. *Eines i feines al camp a Catalunya. L'estudi de l'agricultura a través de l'arqueologia*, 84-88.

Gibaja, J.F., Palomo, A., Terradas, X. and Clop, X. 2004. Útiles de siega en contextos funerarios del 3500-1500 cal ANE en el Noreste de la Península ibérica: El caso de las grandes láminas de sílex. *Cypsela* 15, 187-195.

Gibaja, J.F., Palomo, A. and Terradas, X. 2005. Producción y uso del utillaje lítico durante el mesolítico y neolítico en el Noreste de la Península ibérica. *III Congreso del Neolítico en la Península Ibérica*, 223-231, Santander.

Gibaja, J.F., Palomo, A., Frances, J. and Majó, T. 2006. Les puntes de sageta de l'hipogeu calcolític del carrer París (Cerdanyola): Caracterització tecnomorfològica y funcional. *Cypsela* 16, 127-133.

González, P., Martín, A. and Mora, R. 1999. *Can Roqueta: Un establiment pagès prehistòric i medieval (Sabadell, Vallès Occidental)*. Excavacions Arqueològiques a Catalunya 16, Generalitat de Catalunya, Barcelona.

Grégoire, S. 2000. *Origine des matières premières des industries lithiques du Paléolithique pyrénéen et méditerranéen. Contribution à la connaissance des aires de circulation humaine.*Thèse de doctorat, Université de Perpignan, Perpignan.

Márquez, B. 2004. Los análisis traceológicos como forma de reconstruir las actividades prehistóricas: el caso de la caza. In E. Baquedano and S. Rubio (eds.), *Miscelanea en homenaje a Emiliano Aguirre*. Arqueología. Zona Arqueológica IV (4), 300-311.

Martín, A. 1979. El yacimiento veraciense de "el Coll" (Llinars del Vallès). *XV Congreso Nacional de Arqueología*, Lugo.

Martín, A. and Mestres, J.S. 2002. Periodització des de la fi del Neolític fins a l'Edat del Bronze a la Catalunya Sud-Pirinenca. Cronologia relativa i absoluta. *XII Col.loqui Internacional d'Arqueologia de Puigcerdà*, 77-130.

Mercadal, O. (ed.) 2003. *La Costa de can Martorell (Dosrius, Maresme). Mort i violència en una comunitat del litoral català durant el tercer mil·lenni aC*. Laietania - Estudis d´Arqueologia i d´Història 14. Museu de Mataró.

Mestres, J.S. 2002. La datació per radiocarboni. In X. Clop and M. Faura (eds.), *El sepulcre megalític de Les Maioles (Rubió, Anoia). Pràctiques funeraries i societat a l'altiplà de Calaf (2000-1600 cal ANE)*. Estrat 7, 167-178, Igualada.

Miret, J.M. 1993. La indústria lítica de la Bòbila Madurell. Campanyes de 1987-1988. *Cypsela* 10, 23-32.

Miret, J.M. and Martín, A. 1998. La industria lítica del

jaciment verazià del Coll (Llinars del Vallès). *Lauro* 15, 5-14.

Morgado, A., Pelegrin, J.,Martínez, G. and Afonso, J.A. 2008. La production de grandes lames dans la Péninsule ibérique (c. IV – III mil. cal. A.C.). In M.H. Dias-Meirinho, V. Léa, K. Gernigon, P. Fouére´, F. Briois and M. Bailly (eds), *Les industries lithiques taillées des IVe et IIIe millénaires en Europe occidentale*, British Archaeological Reports, International Series 1884, 309–330. Oxford, BAR Publishing.

Ortí, F., Rosell, L., Salvany, J.M. and Inglés, M. 1997. Chert in continental evaporates of the Ebro and Calatayud Basins (Spain): distribution and significance. In A. Ramos and M.A. Bustillo (eds.), *Siliceous Rocks and* Cultura, 75-89, Granada, Universidad de Granada.

Palomo, A. and Gibaja, J.F. 2002. Análisis de las puntas del sepulcro calcolítico de la Costa de Can Martorell (Dosrius, El Maresme). In I.Clemente, R. Risch, J.F. Gibaja (eds.), *Análisis funcional. Su aplicación al estudio de sociedades prehistóricas*, British Archaeological Reports, International series 1073, 243-249, Oxford, BAR Publishing.

Palomo, A. and Rodríguez, A. 2003. *Memòria dels treballs arqueològics desnvolupats a Can Roqueta II (Sabadell-Vallès Occidental) 1999/2000.* Sabadell. Unpublished report.

Palomo, A., Terradas, X., Clop, X. and Gibaja J.F. 2004. Primers resultats sobre l'estudi de les grans làmines procedents de contextos funeraris del nord-est de la Península Ibérica, *Revista L'Arjau* 48, 24-27.

Pelegrin, J. and Morgado, A. 2007. Primeras experimentaciones sobre la producción laminar del Neolítico reciente - Edad del Cobre del Sur de la Península ibérica. In M.L. Ramos, J.E. González and J. Baena (eds.), *Arqueología experimental en la Península ibérica. Investigación, didáctica y patrimonio,.* Asociación Española de Arqueología Experimental, 131-139, Santander.

Pericot, L. 1950. *Los sepulcros megalíticos catalanes y la cultura pirenaica.* Barcelona, CSIC.

Plisson, H., Mallet, N., Bocquet, A. and Ramseyer, D. 2002. Utilisation et rôle des outils en silex du Grand-Pressigny dans les villages de Charavines et de Portalban (Néolithique final). *Bulletin de la Société Préhistorique Française* 99 (4), 793-811.

Plisson, H., Bressy, C., Briois, F. and Renault, S. 2006. Les productions laminaires remarquables du midi de la France à la fin du Néolithique : les bases d'un programme collectif de recherche. In J. Vaquer, J. and F. Briois (eds.), *La fin de l'Âge de Pierre en Europe du Sud.* Archives d'Écologie Préhistorique,71-83, Toulouse.

Renault, S. 1998. Economie de la matière première. L'exemple de la production au Néolithique final en Provence des grandes lames en silex zoné oligocène du bassin de Porcalquier (Alpes de Haute Provence). In A. D'Anna and D. Binder (eds.), *Production et identité culturelle. Actualité de la Recherche. Rencontres de Préhistoire récente*, 145-161, Antibes, Éditions APDCA.

Renault, S. 2006. La production des grandes lames au Néolithique final en Provence : matériaux exploités, multiplicité des productions, aspects technologiques et chrono-culturels. In J. Vaquer and F. Briois (eds.), *La fin de l'Âge de Pierre en Europe du Sud.* Archives d'Écologie Préhistorique, 139-164, Toulouse.

Roig, J., Molina, D., Coll, J.M. and Molina, J.A. in press. El jaciment calcolític del Vapor Gorina (Sabadell). *Tribuna d'Arqueologia 2007.* Generalitat de Catalunya, Barcelona.

Rovira, C. 2003. Can Roqueta II(Sabadell-Vallès Occidental). Els materials de caire metàl·lic i metal·lúrgic. In A. Palomo and A. Rodríguez (eds.), *Memòria dels treballs arqueològics desnvolupats a Can Roqueta II (Sabadell-Vallès Occidental) 1999/2000.* Sabadell. Unpublished report.

Terradas, X. and Borrell, F. 2002. Les restes lítiques tallades. In G. Alcalde, M. Molist and M. Saña (eds.), *Procés d'ocupació de la Bauma del Serrat del Pont (La Garrotxa) entre 5480 i 2900 cal AC.* Publicacions Eventuals d'Arqueologia de la Garrotxa 7, 30-35, Olot.

Terradas, X., Palomo, A., Clop, X. and Gibaja, J.F. 2005. Primeros resultados sobre el estudio de grandes láminas procedentes de contextos funerarios del Noreste de la Península ibérica. *III Congreso del Neolítico en la Península Ibérica*, 349-357, Santander.

Vegas, J.I. 1999. *San Juan Ante Portam Latinam.* Álava: Exposiciones del Museo de Arqueología de Álava. Diputación Foral de Álava.

Vilaseca, S. 1973. *Reus y su entorno en la prehistoria.* Asociación de Estudios Reusenses, Rues.

A View From the Mines.
Flint Exploitation in the Gargano (South-Eastern Italy) and Socio-Economic Aspects of Raw Materials Procurement at the Dawn of Metal Production

Massimo Tarantini

Dip. di Archeologia e Storia delle Arti,
Sezione di Preistoria – Università di Siena (Italy)
massimotarantini@hotmail.com

Introduction

The establishment of copper working increasingly appears to be the outcome of a series of transformations in the economic, social and symbolic structure of late Neolithic societies leading to a transition to a new phase of history[1]. In itself, the introduction of copper did not cause any significant changes in the day-to-day tools used by Neolithic populations. Stone tools continued to play a vital role for copper age communities. The few functional studies hitherto undertaken on stone tools from this period show clearly that lithic artefacts were not excluded from any sector of production (e.g. Conati Barbaro *et al*. 2002). Copper only replaced stone for the manufacture of some prestige goods (Petrequin *et al*. 2002).

The structural changes marking the transition from the Neolithic to the Copper Age, however, were not without consequences for the production methods employed in the lithic industry. Typological studies and the reconstruction of operational sequences indicate that lithic production was organized in a new way. Whilst it is true that in the Neolithic there is a clear difference between expedient and curated tools (Binder *et al*. 1990), this dichotomy appears to become more pronounced in the Copper Age. Specifically, while expedient tools continue to be produced, prestige objects such as flint knives increase significantly in quality and importance (e.g. Manolakakis 1996; Bailly 2002); at the same time we see the large-scale and standardized production of specific preforms (ogival, bifacially retouched and with a bi-convex cross-section) which later have a wide circulation (e.g. Forenbaher 1999; Campana and Negrino 2002).

The case study proposed here, that of the flint-mines of the Gargano peninsula in south-eastern Italy, shows that changes in the status of lithic industries also went hand in hand, at least in this instance, with a new organization of extraction activities. The economy of a society is an integrated system, whose various components mutually affect one another. Just over 150 years ago, Karl Marx wrote: "production, distribution, exchange and consumption [...] are elements of a totality, differences within a unity: [...] *there is an interaction between the different components*" (Marx 1857-58; tr.it.: 23-24). Relating these different components to one another and attempting to understand how they interact is the key proposed here, as a preliminary working hypothesis, to interpreting the changes observed in the Gargano mines at the transition from the Neolithic to the Copper Age[2].

The Geological and Chronological Context

The Gargano is without doubt the most important area of south Italy as concerns flint supply in recent prehistory. It is a large promontory stretching into the Adriatic sea in south-eastern Italy. Its rock formations are represented principally by various series of carbonate deposits formed during the Jurassic-Cretaceous and middle Eocene. The formations richest in flint, towards which intensive mining activities were directed, are the Maiolica, the Scaglia, and the Nummulite Platform of the Peschici Formation; the latter presents a weak tectonic deformation and the flint contained within it is therefore often found in a virtually undisturbed state.

Hitherto, 18 sites for the extraction of flint have been identified, discovered with their original entrances or during earthmoving for construction or road works. All of these sites can be defined as *mining complexes* as they consist of a number of mines in close proximity; where isolated structures are observed this appears to be due to our partial view of the context.

Many mines still present problems of chronological and/ or cultural attribution; however, in recent years their chronological and/or cultural context has been defined more

---

[1]   This emerges clearly from two recent surveys of the Europe-wide situation: Lichardus-Itten 2007; Strahm 2007. As concerns changes in social organization, see Coudart *et al.* 1999.

[2]   This contribution represents a further reflection on archaeological data already presented in Tarantini 2007, 2008.

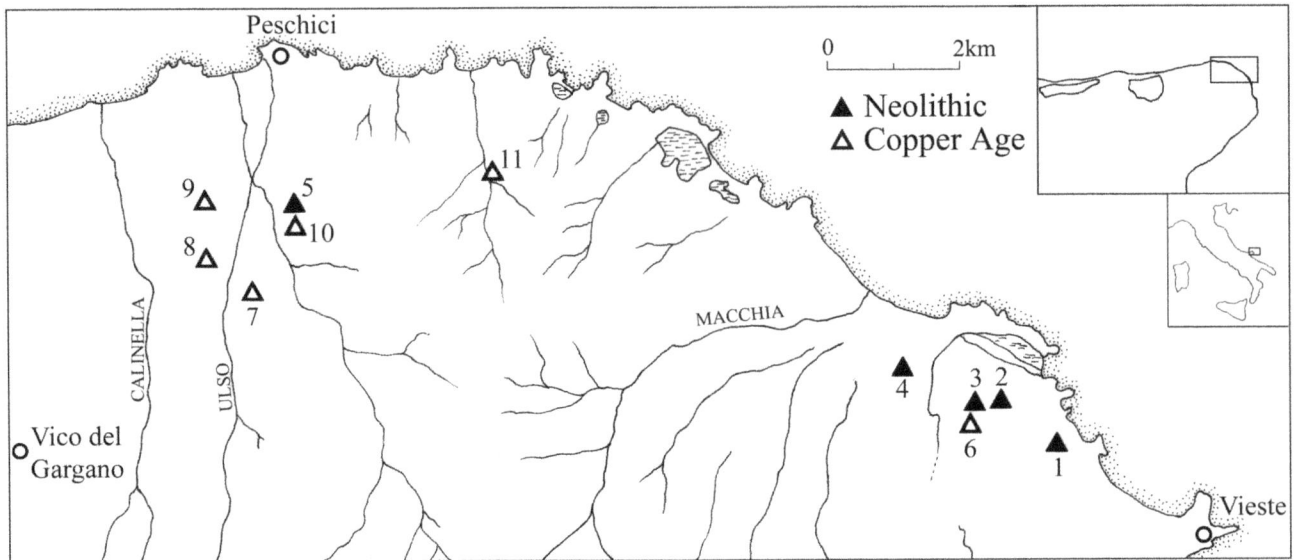

Figure 1: Chronological or cultural data available for the Gargano prehistoric flint mines.

precisely. Eleven complexes have now been attributed to specific contexts (Figures 1, 2), the implications of which will be discussed below. We note here that it is now possible to date the beginnings of extraction activities to the early VI millennium BC (Defensola A: Utc 1342 6990±80 BP, cal. 2σ 5993-5652 BC), coinciding with the Neolithization of south-eastern Italy; the mining phenomenon terminates in the final phase of the Copper Age (Defensola B: Beta 171597 4050±40 BP, cal. 2σ 2850-2820, 2670-2470 BC).

Mining techniques and typologies

Generally speaking, despite the impossibility of determining the chronology of some mines, two distinct types of mining technique appear to have been employed in the Gargano; these appear to be closely linked to the geo-morphological properties of the formations concerned.

1. Sub-horizontal mining. In compact formations, mining proceeded with the removal of individual limestone seams in their entirety, extracting the flint using two different methodologies: "by collapse" or "in steps" (Figure 3). Depending on the thickness of the seams, a sufficient number of these were removed to obtain a floor-ceiling height such as to allow mining to proceed. This height rarely exceeds 60 cm and as such presents considerable logistical difficulties. This excavation technique used the joins between rock layers to guide the removal of the seams, thereby creating structures with a characteristic flat roof.

Mines created using sub-horizontal excavation generally open onto slopes with a convex morphology, and have sometimes produced extremely large conoidal spoil heaps outside. The depth of excavation towards the interior of the hill ranges from about ten metres in some complexes to over 100 metres in Defensola A, and presents a "chamber and pillar" layout: in other words, the mines consist of one

or more communicating chambers, containing pillars of untouched rock and/or heaps of excavation debris, probably used as support structures. The underground surface area may be extremely large; a conservative estimate gives a surface area of at least 6500 m² for Defensola A.

This extraction technique, characterized by a considerable degree of technological skill and involving the repeated application of a single module, is seen exclusively in the mines of the Vieste area, opened into nummulite limestones; it appears to be characteristic of the ancient and early-middle Neolithic (Defensola A and C, S. Marco, Arciprete: ca. 5900-5200 BC).

2. Vertical excavations. Vertical access to flint-bearing formations is the only option in flat terrain, as in large areas of central Europe. Extraction using vertical shafts is also the only option in formations which are heavily affected by tectonic activity or which are crumbly, where it would therefore be impossible to carry out deep sub-horizontal mining safely. Less information is available about this extraction technique, due in part to the difficulty of interpreting structures which are visible mainly in artificial sections (fig. 4.2). However, a comparison with other mining contexts (e.g. Di Lernia and Galiberti 1993; Weisgerber, Slotta and Weiner 1999) has allowed us to identify both "simple" shafts and "access" shafts to subterranean chambers with a sub-horizontal development and limited size; however, we are rarely able to determine their morphological layout (fig. 4.1, 4.3).

This exploitation strategy involved digging multiple shafts, more or less adjacent to one another and giving access to subterranean chambers with limited sub-horizontal development. In contrast to many European mine contexts, the Gargano shafts are rarely of significant depth, on average three metres. The debris from these vertical excavations was

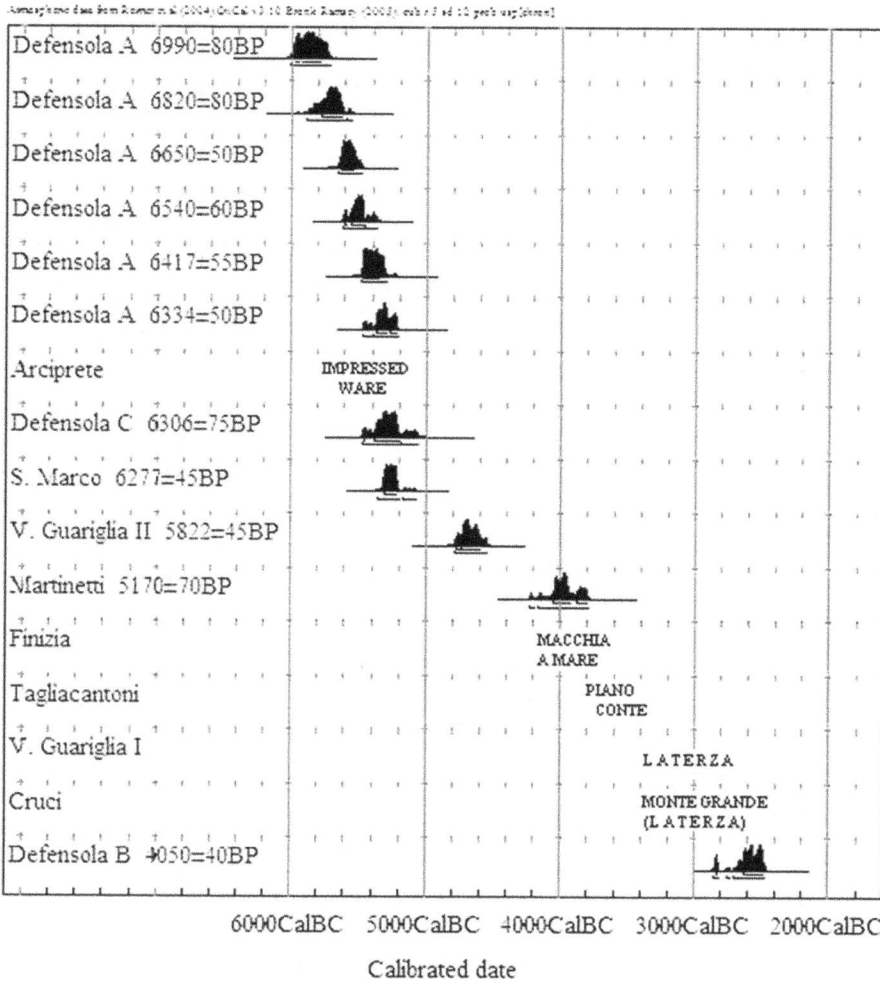

Figure 2: Location map of mines.

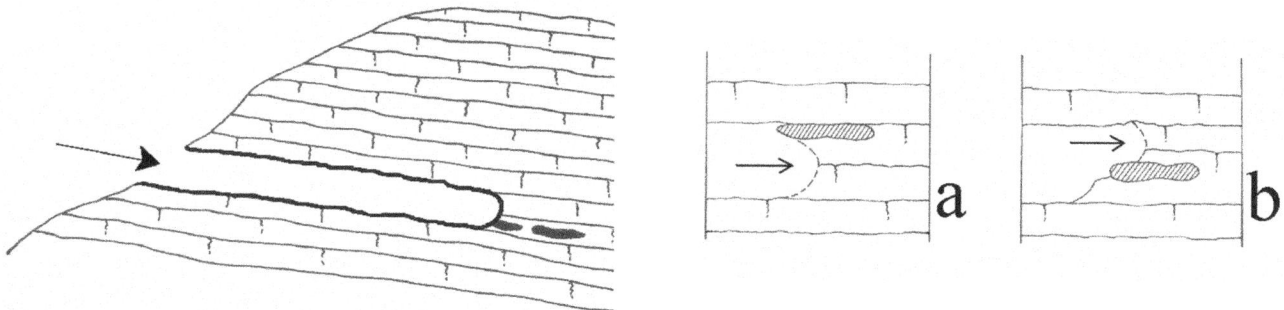

Figure 3: Typology of extraction structures in compact formations (drawing by M. Tarantini) and extraction techniques for flint nodules in compact formations: "by collapse" (a) and "in steps" (b) (from Di Lernia and Galiberti 1993, modified).

partly used to fill earlier shafts or structures, and partly distributed over the land surface in ways only observed in road sections. The area covered by these complexes is difficult to determine with surface surveys; in those rare cases where we have been able to gain a less partial view of the overall context, we are dealing with linear distances of between 200 and 300 metres.

According to the information currently available, this mining technique is documented in the Maiolica and Scaglia formations; it is chronologically restricted to the Copper Age (ca. 4000-2500 BC), with only one and hitherto

isolated event (Valle Guariglia II) dated to the Middle Neolithic (Galiberti and Tarantini 2006). .

Technical changes and the organization of work at the dawn of the Copper Age

The different mining techniques adopted in the Gargano appear to be closely linked to geo-morphological factors and particularly the compactness of the rock formations. The study of the geological context is always a requisite first step in gaining an understanding of the typology of

55

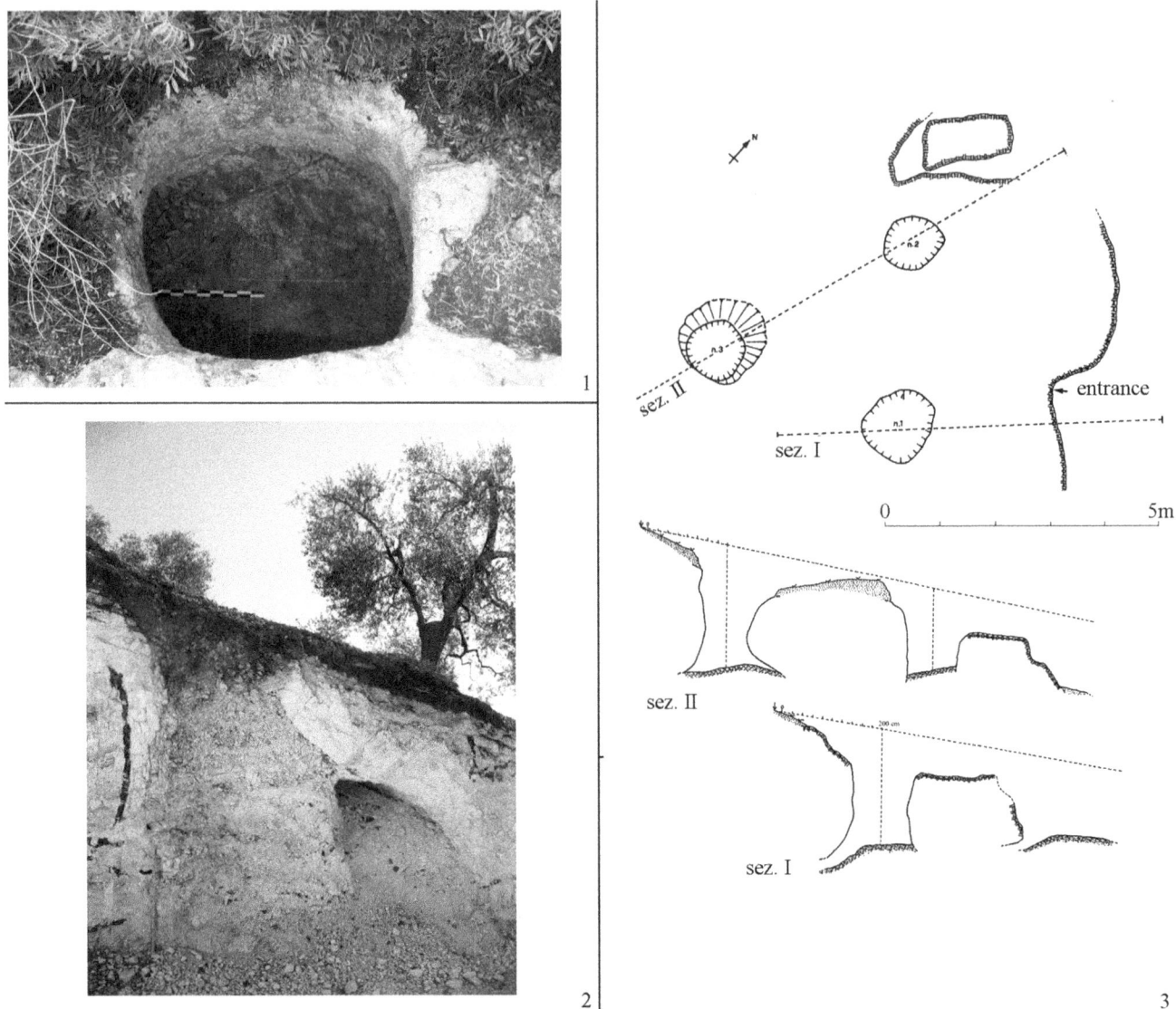

Figure 4: Vertical shaft mining system. (1. Defensola B shaft, still open; 2. Carmine shaft, near Mattinata, discovered as a consequence of earthmoving for construction; 3. Finizia shafts map) (photos by M. Tarantini, drawings by A. Galiberti).

mining structures[3]. Nevertheless, the fact that only sub-horizontal mines in compact formations are attested during the ancient and early-middle Neolithic whereas vertical shafts and mines dug into formations affected by tectonic activity belong exclusively to the Copper Age, clearly suggests that social and economic factors also played a part, leading to a given extraction technique being preferred at a given time.

It is important to stress that the interpretation proposed below is currently a working hypothesis, awaiting further confirmation from a better knowledge of the more wide archaeological context. The Gargano has a long tradition of typological research on the numerous lithic complexes

of the Holocene, contemporary with mining activities and largely linked to these (e.g. Palma Di Cesnola and Vigliardi, 1984; Palma Di Cesnola 1987; Calattini and Cuda 1988); still lacking, however, are studies of lithic technology -the only ones that can help us to know the aims of the production- and systematic excavations and studies of settlements and cemeteries. Despite these gaps in our knowledge, the data on mining structures nonetheless allow us to formulate some working hypotheses. The long-term nature of mining activities makes it possible to see clearly some changes taking place with the transition from the Neolithic to the Copper Age and to evaluate some specific features of mining activities at the dawn of metal working.

The establishment of underground mines for the extraction of flint coincides with the Neolithization of the Gargano in the early VI millennium BC and poses some important questions regarding the modes of production and social organization specific to the early Neolithic. These mines are large structures, with internal galleries well over 100 metres

---

[3]   A typical example is the large mining complex at Krzemionki in Poland where, depending on the depth of the mine, we find first incoherent, then tectonized and then compact rock; the mining techniques used differ depending on the layer affected, clearly in relation to the static potential offered by the surrounding rock (Borkowski 1997). Strong relationships between geological context and excavation technique are observed in the quarries too (La Porta 2005).

long, which are difficult to move around and orientate oneself inside, and which therefore require some form of "initiation". These mines also testify to a high degree of technological sophistication, with clear evidence that they were deliberately planned for long-term exploitation and systematically aimed at deep areas. This is indicated by the digging of large galleries ("fast lanes") cutting through previously accumulated mining debris and aimed at providing more rapid access to the active extraction faces located in the innermost parts of the mine. The fact that some of these galleries, whose construction cannot be underestimated in terms of the manpower required, are bounded along most of their length by dry stone walls on both sides confirms the theory of the planned management of the mine (Galiberti and Tarantini 2005, 200).

Whilst it would be almost tautological to claim the existence of specialized miners, the problem of defining the role played by these specialists in the context of an early Neolithic society remains. On the basis of archaeological data from contemporary contexts in central and southern Italy, we can rule out the existence of social hierarchies. As such, the concept which seems best able to describe the role of these miners is that of *part-time specialists*. In any case, as C. Meillassoux (1975, tr. it.: 47) has noted: "having a speciality does not, in this context, imply specialization, that is the exclusive practice by an autonomous productive unit, of a non food-producing activity that implies continuous transfer of subsistence goods to the specialized unit. Practising a speciality does not necessarily imply relinquishing agricultural work". In other words, it is reasonable to assume that mining activities were carried out by men who can appropriately be described as specialists for the technological skills possessed, but who worked on an occasional and irregular basis, plausibly in accordance with the cyclical organization of production activities characteristic of farming societies. It is important to stress that all known ethnographical data on New Guinean and Australian societies with a comparable level of technology and socio-economic organization agree on suggesting that mining activities are never continuous in nature[4]. This is further confirmed by the experiments conducted on lithic objects manufactured by highly skilled workers, such as that carried out at Grand Pressigny (Pelegrin 2002, 134).

The observations conducted on the Defensola mine also make it possible to draw a parallel between the working methods suggested for the Neolithic mining structures of the Gargano and one of the two trends identified by the ethno-archaeological research of P. and A.M. Petrequin in Irian Jaya. Here the skills required for extraction activities (and the corresponding ritual) are complex, in order to ensure that these skills are passed on within the groups of specialists (Petrequin and Petrequin 2002).

The large complexes of the ancient and early-middle

Neolithic are followed by a period roughly coinciding with the V millennium BC for which we have little evidence of mining. This leads us to suggest a decline in the intensity of mining activities, probably due to local population dynamics which are also clearly reflected in the simultaneous depopulation of the nearby Tavoliere area, where the cycle of large Neolithic ditched villages comes to an end.

A resumption of mining activities is documented at the beginning of the IV millennium, contemporary with the establishment in the Gargano of the Macchia a Mare facies marking the connection between the end of the Neolithic and the beginning of the Copper Age. Mining now takes on very different connotations which become more marked at the height of the Copper Age: this period sees the establishment of structures giving vertical access to the flint-bearing formations, generally small in size and apparently unplanned, certainly requiring a far more limited investment of skills with respect to that needed by Neolithic structures. Unlike the Neolithic mines, where a single structure shows continuous use over several centuries, during the Copper Age there may be continuity in a given area but not in a given extraction structure; these mines generally open into heavily tectonized or crumbly formations where sub-horizontal excavation would have been impossible for safety reasons (Tarantini 2008). The precarious static conditions of these mines suggest that mining was undertaken rapidly and that each shaft therefore represents a single extraction event (see also Georges 1995; Bostyn and Lanchon 1992). Exemplary from this point of view is the Valle Guariglia I mine, a shaft mine filled with debris (and reused as a burial place); after being emptied during archaeological excavation work this shaft suffered significant structural collapse over the course of a few years.

This different extraction technique thus required a much lower degree of know-how than that of the previous period and probably also reflects a different organization of mining work: no longer, as in the Neolithic, a group of people with highly specialized skills who, precisely because of these skills, must have maintained some sort of exclusive right of access to flint-bearing formations[5]. Rather – once again with reference to the trends identified by the Petrequins in Irian Jaya – we find a situation in which all men may potentially extract flint and where work is organized in a way reminiscent of the collective activities of a group of farmers. The most difficult aspect of the work now lies above all in prospecting, in other words the search for flint seams where these are not visible at the surface and in the assessment of their potential; an indication of how difficult this operation was can be found at the Martinetti site where

---

[4] A review of ethnographic accounts of deep-shaft quarrying in De Grooth 1991: 154, to which we should now add the fundamental monograph on the Irian Jaya quarries (Petrequin and Petrequin 2002).

[5] I prefer to use generic terminology rather than make specific reference to concepts of ownership of resources, quarries and mines, since in this case the ethnographic documentation present a wide variety of situations (see De Grooth 1991, 154; on the problems concerning the identification of property in prehistory, see Earle 2000).

a shaft about 2 m deep did not encounter flint beds and can therefore be interpreted as a failed test shaft[6].

The evolution of the mining techniques described and the resulting organization of work hypothesised show clearly that when analysing societies and their technologies it is also essential to bear in mind processes of contraction and involution; some interpretations of evolution as a progressive increase in complexity turn out, in cases such as these, to be wholly insufficient.

The Gargano also allows us to observe another phenomenon of enormous interest: a shift in the investment of technological skills from one sector (or more accurately one segment) of production to another. Specifically, during the same phase which saw the establishment of this new organization of mining work, the Gargano saw a general typological and structural modification of lithic production. Compared to the Neolithic we see a greater typological variety, a higher overall degree of laminarity, the emergence of flaked artefacts – which later become established as one of the most important typologies – and a general attention to profiling in fairly commonplace objects like tranchets (Palma Di Cesnola 1987). The extraction of flint now appears to be practicable by all, at least in terms of the technical skills needed, whilst specialist skills seems to shift towards *débitage*.

The shift of technological skills in the direction noted above seems to be confirmed by the spatial organization of production. During the Copper Age, at least as far as we can judge from data from the Ulso valley near Peschici, lithic workshops are systematically documented. In other words, in some cases there seems to be a segmentation of the operational sequence with a distinction between the place of extraction and the place where the supports were transformed, which may be considered indicative of an increasing degree of know-how (Petrequin and Jeunesse 1995, 69). It is worth noting that the same distinction between extraction areas and working areas has been observed in three Copper Age jasper quarries found in the Apennines between Liguria and Emilia Romagna (Ghiretti *et al.* 2002, 406).

Flint mines as part of an integrated system

At the current state of knowledge it is extremely difficult to understand the reasons underlying the change in mining techniques and typologies in the Gargano during the IV millennium BC. The establishment of the Macchia a Mare facies seems to have affected the northern part of the Gargano, from Punta Manaccore just east of Peschici to Lake Varano. As such, this factor may lie behind the enlargement of the area where mining activities took place and which in the Neolithic was restricted to the environs of Vieste. This enlargement led to the exploitation of the less compact limestone formations in the Peschici area and

this in turn may explain the new types of mines observed. However, it should also be noted that even in the Vieste area (Defensola B mines: Tarantini 2003) the late Copper Age saw the exploitation of less compact detritic formations (Figure 5). It is obvious that we are dealing with a series of deliberate choices which make it possible to rule out any form of strict environmental determinism.

At this point it is worth checking whether the developments observed in the Gargano can be correlated with more widespread phenomena identified at the transition from the Neolithic to the Copper Age in Italy and elsewhere in Europe and if it is therefore reasonable to suggest that the changes observed in mining activities in the Gargano form part of structural changes with a much broader impact.

The element that marks the most evident change in the Copper age lithic industries is the copious and systematic presence of arrowheads, or more generically of "bifacial heads". It is too early to say if this instruments were used for hunting or war (arguments in Lemorini and Massussi 2003); anyway, their frequent presence in burial contexts, as for instance in those of the Gaudo in Campania and southern Lazio, seems to indicate their connotative value of the male identity.

However, what interests more here is that the only context of siliceous raw material supplying studied in systematic way in Italy, Valle Lagorara in Liguria, has pointed out as the extraction and the first working realized *in situ* were oriented to ogival preforms from which to obtain arrowheads (Campana and Maggi 2002). A similar *chaîne opératoire* has been evidenced in another context rich in raw materials like the Marche, in particular thanks to the re-examination of the materials coming from the ditch of Conelle di Arcevia (cfr. Albertini 2003). Therefore it does not seem daring to assume -even if in the total absence of technological studies- that at least one part of the Gargano Copper Age bifacial component is to be connected to the preparation of preforms for arrowheads; an indication in this direction it may be stated for instance from a simple look to the tables connected to the publication of the materials of Pagliara di Malanotte (Calattini and Cuda 1984) (Figure 6). For the lack of technological studies already pointed out, the problem of the spread of bifacial preforms in southern Italy has not been faced at the moment; however the study led from A. Musacchio on the lithic component of the Late-Neolithic site of Quadrato di Torre Spaccata, has clearly shown as the arrowheads found in this site are obtained *in situ* from bifacial preforms (Musacchio 1995). Moreover, experimental studies suggest that the production of preforms is the most delicate phase, whereas the transformation of a preform in an harrowhead led to pratically no accidents but require about an hour, instead of 5-30 minutes required from the production of a preform (Briois and Negrino 2002). Besides, to produce preforms in the same supplying place allows to carry a less quantity of raw-material. Generally speaking it was therefore more

---

[6]   Another example of the digging of "*puits test*" can be found in De Labriffe and Thèbault 1995.

0   5m

Figure 5: Defensola B, entrance B. Section and inner view (drawing and photo by M. Tarantini).

convenient to realize the preforms in proximity of the places of raw materials supplying.

Moreover, it is important to note that the resumption of mining activities in the Gargano at the beginning of the IV millennium coincides roughly with the interruption of some long-distance distribution networks for raw materials which had characterized the Italian Neolithic (obsidian: Bigazzi and Radi 1996; Tykot 1996; Vaquer 2006; greenstones: D'Amico 2000; flint itself in northern Italy: Ferrari *et al.* 1998). More than by the introduction of new raw materials, which will become important much later, the drastic reduction in the long-distance Neolithic trade networks may be indicative of the collapse of that network of alliances which made them possible or of a structural changes in exchange system.

In the same period, the pottery record shows a process of gradual disintegration of the homogeneity of the final Neolithic, in line with a trend characterizing the transition from the Neolithic to the Copper Age throughout Europe (Whittle 1996). Despite some strong elements of continuity, the Adriatic regions of central and southern Italy also saw the emergence of autonomous and more circumscribed facies: this is true of Abruzzo, where Ripoli evolved into three distinct descendant facies (Ripoli III, Paterno, Fossacesia: Pessina and Radi 2003), and of Puglia, where Diana was also succeeded by descendant facies such as Macchia a Mare and Zinzulusa (Cremonesi 1984; Cazzella 1992).

In direct correlation with this processes, we observe an alteration in the relationship between individual populations

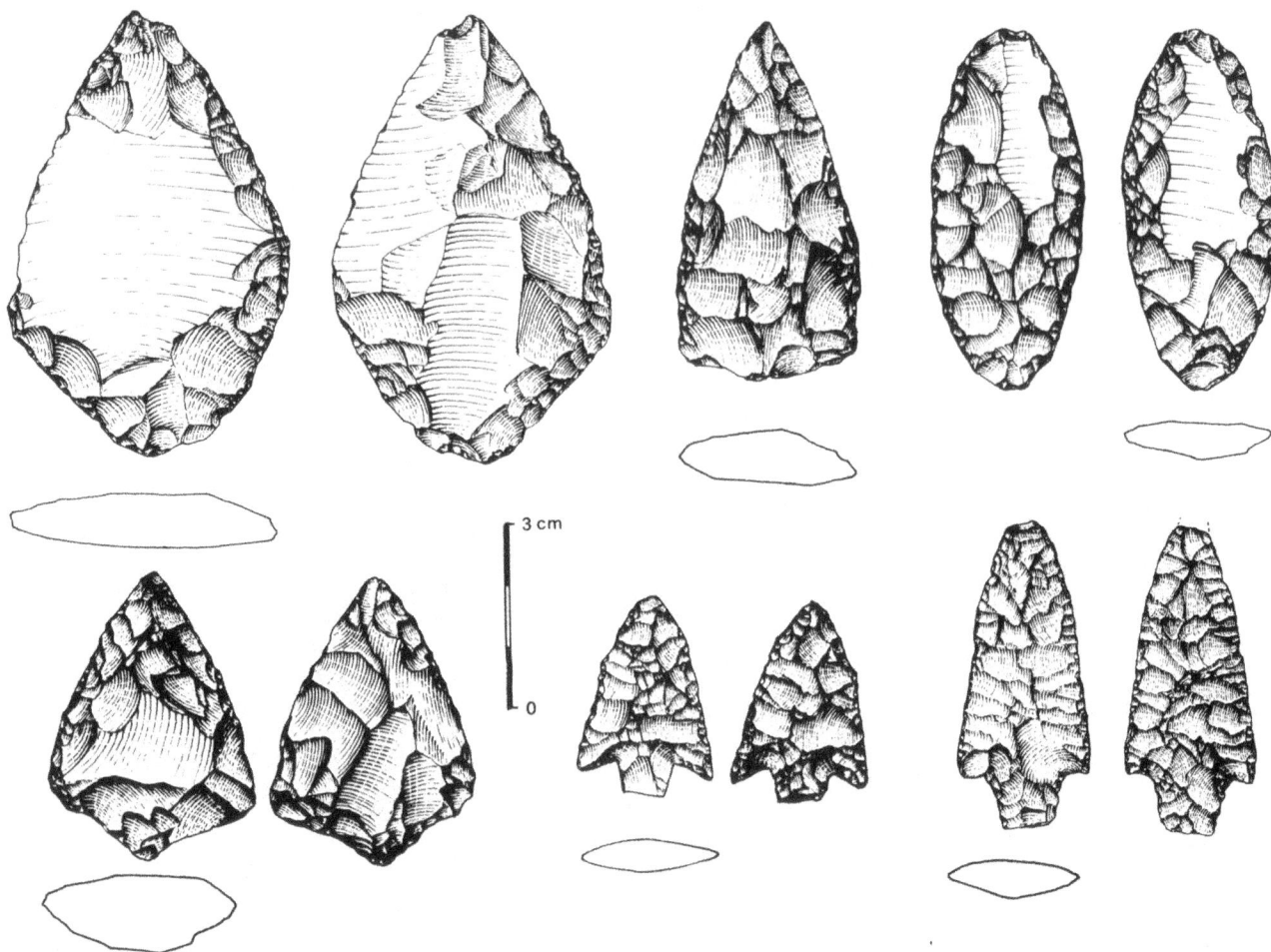

Figure 6: Lithics from Pagliara di Malanotte (Peschici, FG) (from Calattini, Cuda 1987, modified).

and their territory. This is clearly shown by the systematic exploitation during this period of local outcrops of lithic raw materials, even when these are of inferior quality. These are widespread processes observed on a European level at this time (Guilaine 1994, 88-89) and Italy is no exception, although only the central and northern parts of the peninsula are well understood. For example, in Emilia Romagna and Friuli, so-called "alpine" flint is still attested, but for the production of objects such as daggers, which undergo a significant development precisely during the Copper Age (Mottes 2001); at the same time we see a "more systematic exploitation of local resources, such as Apennine materials in Emilia, Holocene beach pebbles in Romagna and quartzites in Friuli" (Ferrari *et al.* 1998, 16).

Throughout northern Italy the most significant evidence of trade becomes that relating to metals, used to create prestige objects, and ornaments such as necklaces in marble, steatite or seashells (Barfield 1981). Tuscany also sees a significant broadening of the sources of supply for different types of materials: this period saw the opening of jasper quarries (Gambassini and Marroni 1998), the beginnings of the extraction of metals and cinnabar and the exploitation of marble, quartz and steatite, materials which appear to have been sought out to make ornamental objects

(Grifoni Cremonesi 1988-89; Cocchi Genick and Grifoni Cremonesi 1991). Similarly, in Liguria we find evidence for copper mines at Libiola and Monte Loreto (Campana *et al.* 1998; Maggi and Pearce 2005), for a steatite worksite at Pianaccia di Suvero (Maggi and Del Lucchese 1988) and for jasper quarries at Valle Lagorara (Campana and Maggi 2002) and Boschi di Liciorno (Campana *et al.* 1998). Generally speaking, in Liguria there is a drastic reduction in the use of raw materials from other areas and a preference for the use of lithotypes found within a radius of a few tens of kilometres (Negrino and Starnini 2006, 293). The use of local flint, even if of lower quality, characterise the Copper age flint industry of Monte Covolo, in Lombardia, differently from the Neolithic industry (Lo Vetro 2002, 258).

The examples provided raise another issue: the decline in trade seems to involve the raw materials used to produce objects for domestic use, whilst at the same time the production of ornamental and prestige objects, which circulated over long distances, seems to increase. In other words, trade networks are not interrupted but change in nature.

Lithic production during the Copper Age is an integral part

of this production process of prestige items. The production of prestige objects in flint has been clearly documented in northern Italy, where over 400 flint daggers are been identified (Mottes 2001). Southern Italy is still little studied from this point of view. But the flint daggers, sometimes more than 20 cm. long, are a recurrent element of the Gaudo burials in Campania (e.g. S. Antonio di Buccino: Holloway 1973: Plates XI, XIII, XVII, XIX-XXII; Pontecagnano: Bailo Modesti and Salerno 1998: Plates 53-63) and in southern Lazio (Carboni 2002); it is important to point out that both Campania and Lazio are areas lacking in important flint sources and therefore -above all for Campania- it has been suggested an origin of flint from the Gargano. After all, the production of these artefacts demands cores of big size. These may only come from lists or nodules in primary position. In Puglia grave goods have not artifacts such as those present in Gaudo burials, but the hoards of Altamura and Monteparano in the central Puglia, containing numerous flint daggers (Peroni 1967), confirm the value of these objects also in an area in which they are not used as grave goods.

Data reported, though limited, seem sufficient to point out that the resumption of mining and the technological changes in flint extraction activities in the Gargano form part of a large-scale processes that involve at the same time lithic technology and tipology, the social *status* of some flint instruments and the mechanism of exchange.

References

Albertini, D. 2003. Tecnologia della lavorazione bifacciale. In A. Cazzella, M. Moscoloni and G. Recchia, *Conelle di Arcevia. II**. I manufatti in pietra scheggiata e levigata, in materia dura di origine animale, in ceramica non vascolari; il concotto*, 1-92. Roma, Casa Editrice Università La Sapienza/Rubbettino.

Bailly, M. 2002. Du Néolitihique final à l'age du Bronze ancien en Bassin Rhodanien. Une première approche du statut des production lithiques. In M. Bailly, R. Furestier and T. Perrin (eds.), *Les industries lithiques taillées holocènes du Bassin rhodanien: problèmes et actualités*, 205-223. Montagnac.

Bailo Modesti, G. and A. Salerno 1998. *Pontecagnano. II.5. La necropoli eneolitica. L'età del Rame in Campania nei villaggi dei morti*, Napoli, Istituto Universitario Orientale, Quaderno n. 11.

Barfield L. H. 1981. Patterns of north italian trade 5000-2000 BC. In G. Barker and R. Hodges (eds.), *Archaeology and Italian society*, 27-51, British Archaeological Reports IS 102. Oxford.

Bigazzi, G. and G. Radi 1996. Prehistoric exploitation of obsidian for tool making in the Italian peninsula: a picture from a rich fission-track data-set, *XIII International Congress of Prehistoric and Protohistoric Science*, Proceedings, 1, 149-156. Forlì.

Binder D., Perlès C., Inizian. M.-L. and M. Lechevallier 1990. Stratégies de gestion des outillages lithiques au Néolithique, *Paléo* 2, 257-283.

Borkowski, W. 1997. Exploitation field in Krzemionki. System of deposit utilisation. In A. Ramos Millan and M. A. Bustillo (eds.), *Sicileous rocks and culture, VI International Flint Symposium (Madrid 1-4 October 1991)*, 327-335, Granada.

Bosellini, A. and M. Morsilli 2002. *Il Promontorio del Gargano. Cenni di geologia e itinerari geologici*, s.l.

Bostyn, F. and Y. Lanchon (eds.) 1992. *Jablines. Le Haut Château (seine-et-Marne). Une minière de silex au Néolithique*, Paris, Documents d'archéologie française, n. 35.

Briois F. and F. Negrino 2002. Riproduzione e verifica sperimentale della catena operativa di Valle Lagorara. In N. Campana and R. Maggi (eds.) *Archeologia in Valle Lagorara. Diecimila anni di storia intorno a una cava di diaspro*, 219-233. Firenze, Istituto Italiano di Preistoria e Protostoria.

Calattini, M. and M. T. Cuda 1984. La stazione di Pagliara di Malanotte in Comune di Peschici: l'industria litica, *Atti 5° Convegno Nazionale sulla Preistoria, Protostoria, Storia della Daunia* (S. Severo 1983), 161-188 and pl. XLII-L.

Calattini, M. and M. T. Cuda 1988. L'Eneolitico del Gargano: l'industria litica e la sua evoluzione. In *L'età del Rame in Europa* (Atti del Congresso internazionale, Viareggio 1987), *Rassegna di Archeologia*, 7, 563.

Campana, N., Maggi, R. and F. Negrino 1998. Le cave di diaspro di Valle Lagorara e Boschi di Liciorno. In A. Del Lucchese and R. Maggi (eds.), *Dal diaspro al bronzo. L'età del Rame e l'età del Bronzo in Liguria*, 145-147. La Spezia.

Campana, N. and R. Maggi (eds.) 2002. *Archeologia in Valle Lagorara. Diecimila anni di storia intorno a una cava di diaspro*, Firenze, Istituto Italiano di Preistoria e Protostoria.

Campana, N. and F. Negrino 2002. L'industria litica scheggiata: tipologia e tipometria. In Campana, N. and R. Maggi (eds.), *Archeologia in Valle Lagorara. Diecimila anni di storia intorno a una cava di diaspro*, 137-211. Firenze, Istituto Italiano di Preistoria e Protostoria.

Carboni, G. 2002. Territorio aperto o di frontiera? Nuove prospettive di ricerca per lo studio della distribuzione spaziale delle facies del Gaudo e di Rinaldone nel Lazio centro-meridionale, *Origini* XXIV, 235-301.

Cazzella, A. 1992. Sviluppi culturali eneolitici nella penisola italiana. In A. Cazzella and M. Moscoloni, *Neolitico ed Eneolitico*, Popoli e civiltà dell'Italia antica, 11, 351-594. Bologna, Biblioteca di Storia Patria.

Cocchi Genick, D. and R. Grifoni Cremonesi 1991. Osservazioni sulle attività minerarie e metallurgiche nel Calcolitico italiano. In P. Ambert (ed.), *Le Chalcolithique en Languedoc et ses relations extra-regionales*, Montpellier 1991.

Conati Barbaro, C., Lemorini, C. and A. Ciarico 2002. Osservazioni sul potenziale interpretativo delle industrie litiche: un'applicazione a contesti del Neolitico tardo in Italia centrale. In A. Ferrari and P. Visentini (eds.), *Il declino del mondo neolitico*, Quaderni Museo Archeologico del Friuli occidentale 4, 167-176

Coudart, A., Manolakakis, L. and J.P. Demoule 1999. Égalité et inégalité sociales en Europe aux VIe et V^e millénaires avant notre ère. In P. Descola, J. Hamel, P. Lemonnier (eds.), *La production du social. Autour de Maurice Godelier*, 267-288. Paris, Fayard.

Cremonesi, G. 1984. Osservazioni su alcuni aspetti dell'Eneolitico del versante adriatico, *Atti 3° Convegno Nazionale Preistoria Protostoria e Storia della Daunia*, 131-147 and plates XXXVIII-XL.

D'Amico, C. 2000. La pietra levigata neolitica in Italia settentrionale e in Europa. Litologia, produzione e circolazione. In A. Pessina and G. Muscio (eds.), *La neolitizzazione tra Oriente e Occidente*, 67-80. Udine.

De Grooth, M.E.T. 1991. Socio-economic aspects of Neolithic flint mining: a preliminary study, *Helinium* XXXI/2, 153-189.

De Labriffe, P. A. and D. Thébault 1995. Mines de silex et grands travaux, l'autoroute A5 et les sites d'extraction du Pays d'Othe. In J. Pelegrin and A. Richard (eds.), *Les Mines de silex au Néolithique en Europe. Avancées récentes*, 48-66, Nancy.

Di Lernia, S. and A. Galiberti 1993. *Archeologia mineraria della selce nella preistoria. Definizioni, potenzialità e prospettive della ricerca*, Quaderni Dipartimento di Archeologia e Storia delle Arti, Sezione archeologica – Università di Siena.

Earle, T. 2000. Archaeology, property, and prehistory, *Annual Review of Anthropology* 29, 39-60.

Ferrari, A., Fontana, F., Pessina, A., Steffe', G. and P. Visentini 1998. Provenienza e circolazione delle rocce silicee scheggiate fra Mesolitico ed Età del Rame in Emilia centro-orientale, Romagna e Friuli, *Archeologia dell'Emilia-Romagna* II/1, 13-19.

Forenbaher, S. 1999. *Production and exchange of bifacial flaked stone artifacts during the portuguese Chalcolithic*, British Archaeological Reports IS 756. Oxford.

Galiberti, A. (ed.) 2005. *Defensola. Una miniera di selce di 7000 anni fa*. Siena, Protagon.

Galiberti, A. and M. Tarantini 2005. La Defensola nel quadro dell'archeologia mineraria europea e del Neolitico antico/The Defensola mine in the context of european mining archaeology and the early Neolithic. In A. Galiberti (ed.), *Defensola. Una miniera di selce di 7000 anni fa*, 197-206. Siena, Protagon.

Galiberti, A. and M. Tarantini 2006. Prima nota sulle miniere di selce di Finizia e gli inizi dell'attività estrattiva nel retroterra di Peschici (FG) nell'età del Rame, *Rassegna di Archeologia*, 22A, 145-152.

Gambassini, P. and G.Marroni 1998. Scoperta di una cava preistorica di diaspro in Val di Farma, *Rassegna di Archeologia* 15, 51-54.

Georges, E. 1995. L'exploitation minière a Saint-Mihiel (Meuse). In J. Pelegrin and A. Richard (eds.), *Les mines de silex au Néolithique en Europe. Avancées récentes*, Nancy, 27-45. Comité des Travaux Historiques et Scientifiques, Nancy.

Ghiretti, A., Negrino, F. and C. Tozzi 2002. Estrazione del diaspro produzione di strumenti a ritocco bifacciale in località Ronco del Gatto (Bardi, Parma): modificazioni economiche e tecnologiche tra la fine del Neolitico e l'età del Rame nell'Appennino ligure-emiliano. In A. Ferrari and P. Visentini (eds.), *Il declino del mondo neolitico*, Quaderni Museo Archeologico del Friuli occidentale, 4, 403-408.

Godelier, M. 1977. *Horizon, trajets marxistes en anthropologie*, Paris, Maspero (tr. it., *Antropologia e marxismo*, Roma, Editori Riuniti 1980).

Grifoni Cremonesi, R. 1988-89. Osservazioni sulla ricerca e lo sfruttamento di alcune materie prime nell'Eneolitico toscano, *Origini* XIV, 253-269.

Guilaine, J. 1994. *La mer partagée. La Méditerranée avant l'écriture. 7000-2000 avant Jesus-Christ*, Paris, Hachette.

Holloway, R. R. 1973. *Buccino. The Eneolithic necropolis of S. Antonio and other prehistoric discoveries made in 1968 and 1969 by Brown University*, Roma, De Luca.

La Porta, P. C. 2005. A geological model for the development of bedrock quarries, with an ethnoarchaeological application. In P. Topping and M. Lynott, *The cultural landscape of prehistoric mines,* 123-139. Oxford, Oxbow book.

Lemorini, C. and M. Massussi 2003. Lo studio dei foliati in selce di Conelle di Arcevia: approccio tecno-funzionale, sperimentale e delle tracce d'uso. In A. Cazzella, M. Moscoloni and G. Recchia, *Conelle di Arcevia. II**. I manufatti in pietra scheggiata e levigata, in materia dura di origine animale, in ceramica non vascolari; il concotto*, 309-351. Roma, Casa Editrice Università La Sapienz /Rubbettino.

Lichardus-Itten, M. 2007. Le Chalcolithique: une époque historique de l'Europe. In J. Guilaine (ed.), *Le Chalcolithique et la construction des inégalités. Tome 1, Le continent européen*, 11-22. Paris, Errance.

Lo Vetro, D. 2002. Il Neolitico di Monte Covolo (scavi 1998-1999): osservazioni sulle industrie litiche, *Rivista di Scienze Preistoriche* LII, 231-260.

Maggi, R. 2001. Pastori, miniere, metallurgia nella transizione fra Neolitico ed età del Rame: nuovi dati dalla Liguria. In A. Ferrari and P. Visentini (eds.), *Il declino del mondo neolitico*, Quaderni Museo Archeologico del Friuli occidentale, 4, 437-440.

Maggi, R. and A. Del Lucchese 1988. Aspects of the Copper Age in Liguria. In *L'Età del Rame in Europa*, Atti del Congresso Internazionale (Viareggio 1987), *Rassegna di Archeologia*, 7, 331-338.

Maggi, R. and M. Pearce 2005. Mid fourth-millennium copper mining in Liguria, north-west Italy: the earliest known copper mines in Western Europe, *Antiquity* 79, 303, 66-77.

Manolakakis, L. 1996. Production lithique et émergence de la hiérarchie sociale: l'industrie lithique de l'Énéolithique en Bulgarie (première moitié du IVe millénaire), *Bulletin de la Société Préhistorique Française* 93, 1, 119-123.

Marx, K. 1857-58. *Grundrisse der Kritik der politischen Okonomie*. Rohentwurf, Berlin 1953 (tr. it., Torino, Einaudi 1976)

Meillassoux, C. 1975. *Femmes, greniers & capitaux*. Paris,

Maspero (tr. it., *Donne, granai e capitali*, Bologna, Zanichelli 1978).

Mottes, E. 2001. Bell Beakers and beyond: flint daggers of northern Italy between technology and typology. In F. Nicolis (ed.), *Bell Beakers today. Pottery, people, culture, symbols in prehistoric Europe*, Proceedings of the International colloquium (Riva del Garda 1998), II, 519-545.

Musacchio, A. 1995. L'industria litica dell'area 3 del sito neolitico di Quadrato di Torre Spaccata (Roma): analisi tecnologica della catena operativa delle punte di freccia ritoccate a pressione, *Origini* XIX, 227-251.

Negrino, F. and E. Starnini 2006. Modelli di sfruttamento e circolazione delle materie prime per l'industria litica scheggiata tra Paleolitico inferiore ed età del Rame in Liguria. In *Materie prime e scambi nella preistoria italiana*, Atti XXXIX Riunione Scientifica IIPP, 283-298, Firenze.

Palma Di Cesnola, A. and A. Vigliardi 1984. Il Neo-eneolitico del promontorio del Gargano. In M. Mazzei (ed.), *La Daunia antica*, 55-74. Milano, Electa.

Palma Di Cesnola, A. 1987. Studio sistematico del primo Eneolitico del Gargano. 1. Studi e considerazioni sulla facies di Macchia a Mare, *Atti del 5° Convegno di Studi sulla Preistoria e Storia della Daunia* (San Severo, 1983), 85-113.

Pelegrin, J. 2002. La production des grandes lames de silex du Gran-Pressigny. In J. Guilaine (ed.), *Materiaux, productions, circulations du Néolithique à l'Age du Bronze*, 131-148. Paris, Errance.

Peroni, R. 1967. *Archeologia della Puglia preistorica*. Roma, De Luca.

Pessina, A. and G. Radi 2003. Il Neolitico recente e finale in Abruzzo, Atti XXXVI Riunione Scientifica IIPP, 209-218, Firenze.

Petrequin, P., Cassen, S., Croutsch, C. and M. Errera 2002. La valorisation social des longues haches dans l'Europe néolithique. In J. Guilaine (ed.), *Materiaux, productions, circulations du Néolithique à l'Age du Bronze*, 67-98. Paris, Errance.

Petrequin, P. and C. Jeunesse (eds.) 1995. *La hache de pierre. Carrières vosgiennes et échanges de lames polies pendant le Néolithique (5400-2100 av. J.-C.)*. Paris, Errance.

Petrequin, P. and A. M. Petrequin 2002. *Ecologie d'un outil: le hache de pierre en Irian Jaya (Indonésie)*, Paris, Monographie du CRA, 12.

Puglisi, S. M. 1948. Le culture dei capannicoli sul Promontorio Gargano, *Memorie dell'Accademia dei Lincei, Cl. Sc. Morali, Storiche e Filologiche*, s. VIII, vol. II, f. I

Strahm, C. 2007. L'introduction de la métallurgie en Europe. In J. Guilaine (ed.), *Le Chalcolithique et la construction des inégalités. Tome 1, Le continent européen*, 49-71. Paris, Errance.

Tarantini, M. 2003. Prime ricerche nel complesso minerario della Defensola B (Vieste- FG). *Atti del 23° Convegno Nazionale sulla Preistoria, Protostoria, Storia della Daunia* (S. Severo 2002), 47-58.

Tarantini, M. 2007. Le miniere neolitiche ed eneolitiche del Gargano. Tecniche estrattive e dinamiche diacroniche, Atti XXXIX Riunione Scientifica IIPP, I, 343-353, Firenze.

Tarantini, M. 2008. Changements techniques au IVe millénaire. Les mines de silex au Gargano (Italie) dans le contexte italienne. In M. H. Dias-Meirinho *et al.* (eds.), *Les industries lithiques taillées des IVe et IIIe millénaires en Europe occidentale*, British Archaeological Reports IS 1884, 331-346. Oxford.

Tykot, R. H. 1996. Obsidian procurement and distribution in the central and western Mediterranean, *Journal of Mediterranean Archaeology* IX, 1, 39-82.

Vaquer, J. 2006. La diffusion de l'obsidienne dans le Néolithique de Corse, du Midi de la France et de Catalogne. In *Materie prime e scambi nella preistoria italiana*, Atti XXXIX Riunione Scientifica IIPP, 483-498, Firenze.

Weisgerber, G., Slotta, R., and J. Weiner (eds.) 1999. *5000 Jahre Feuersteinbergbau. Die Suche nach dem Stahl der Steinzeit*. Bochum, Deutsches Bergbau-Museum.

Whittle, A. 1996. *Europe in the Neolithic*. Cambridge, Cambridge University Press.

FUNCTIONAL VARIABILITY AND ACTIVITY AREAS IN PREHISTORIC MINING OPERATIONS IN NORTHERN CHILE

Diego S. Salazar and Hernán W. Salinas

Arqueólogo. Departamento de Antropología
Universidad de Chile, Santiago (Chile)

David D. Órdenes and Jessica N. Parra

Licenciado/a en Ciencias con mención en Física
Universidad de Chile, Santiago (Chile)

Corresponding author, Diego Salazar: dchbc@123.cl

Introduction

The territory that is encompassed by present-day Northern Chile is the southernmost part of the most arid desert in the world, the Atacama desert. The scarcity of water resources and vegetation in this extensive area contrasts with its rich abundance of metallic and non-metallic ores. Within its mineral diversity, copper plays an outstanding role given the wealth of deposits and the importance that its exploitation has had for the economic development of the nation. In fact, Chile is currently the world's largest copper producer and also holds both the largest open cut mine and the operation with the biggest annual production volume in the world.

This great mineral wealth certainly could not go unnoticed for the indigenous populations that have inhabited these territories since at least 13,000 years ago. Indeed, Northern Chile is one of the earliest South American zones where the use of copper has been documented. Judging by evidence available today, the exploitation of copper ores as semiprecious stones began during the Late Archaic Period, about 3500 years BC (Núñez 2006). This stage in the cultural development of the region is characterized by societies of hunter-gatherers of increasing complexity, at advanced stages of animal domestication and undergoing important social, economic and cultural transformations. The use of copper minerals as semiprecious stones for the creation of necklace beads and other body ornaments has been documented at several sites in the region (Soto 2006, Núñez 1992). Perhaps as a result of centuries of use and experimentation with this raw material, around the year 1200 BC –coincident with the beginning of the Formative Period– the metallurgy of copper and gold made its appearance in the Chilean Atacama Desert (Ayala 2001; Núñez 1987, 1994, 1999, 2006; Muñoz 1989; Salazar 2003-2004). From that moment on, metallic artifacts and, by extension, metallurgic activities, begin to play an important role in the processes of social differentiation and complexity of the atacamenian desert societies.

Nonetheless, the appearance of metallurgy during the Formative Period did not imply the end of the lapidary industry. Rather, the exploitation of copper minerals as semiprecious stones initiated in the previous period acquired a special importance and regional development during Formative times (Rees 1999, Rees and De Souza 2004; Soto 2006, Núñez et al. 2006). Therefore, from the Formative Period on, the exploitation of copper minerals had as its objective the supply of both lapidary and metallurgical activities. In later periods, copper minerals begun to be employed also as incrustations in artifacts made of wood, bone or textile, as well as to be used as ritual offerings to local divinities (Berenguer 2004). Also, recent works document the use of copper to produce mineral pigments for some of the rock art of the region (Sepúlveda and Laval 2010). Even in the 17th century we still find evidence of the use of copper minerals as semi-precious stones (Barba 1967 [1640]).

Given the special symbolic importance of copper, both in mineral and metallic form, and the different contexts in which it was used, the exploitation and processing of copper ores became a relevant part of social, economic and political dynamics of the region, especially considering the uneven geographical distribution of these resources (Núñez 1987; González 2004; Lechtman 1980).

Such is the importance of mining and metallurgy in local prehistory that most researchers usually assume that the main motivation to dominate present-day Northern Chile for both the Tiwanaku state (ca. 500 – 1000 AD) and the Inca Empire (ca. 1450 – 1540 AD) was the control of local copper resources (Llagostera 1976; Raffino 1981; Niemeyer and Schiappacasse 1988; Berenguer and Dauelsberg 1989; Adán and Uribe 2005; Núñez 1999 and 2006; Uribe 1999-2000; Berenguer 2004; Salazar 2002-2005, among others). Despite all this, little is yet known about the indigenous mining and metallurgical productive processes throughout the long cultural history of present-day Northern Chile.

Our knowledge is fundamentally based on the analytical study of finished objects, made both in metal and copper ores, but systematic research about the productive processes have been very scarce, especially in the case of mining. Undoubtedly, the above is partly due to the fact that there is little direct evidence of indigenous mining and metallurgical activities that have survived centuries of medium and large-scale exploitation. In fact, after a systematic survey carried out in the early '70s, Heather Lechtman was unable to identify indisputably pre-Hispanic mines in all of the Peruvian territory and the southern Andean altiplano (Lechtman 1976). Though currently indigenous pre-Hispanic mines are known in Perú and Northern Chile (cf. Iribarren 1962, 1971; Petersen 1970; Shimada 1994; Westfall and González 2006), these have not been studied or published in detail, and thus the impression handed down to us by the first Spanish conquistadores about indigenous mining is still much prevalent (Cf. Boman 1908; Latcham 1938; Petersen 1970; Ravines 1978).

In this context, our work in recent years in San José del Abra and Conchi Viejo, in the highlands of Northern Chile (Figure 1), has attempted to establish the archaeological study of mining activities as a research goal in itself. We have therefore focused on the systematic identification and record of direct evidences of pre-Hispanic mining, as well as the formulation of specific research strategies with its own theoretical and methodological frameworks (Salazar 2003-2004; Salazar and Salinas 2005, 2007; Salazar et al. 2005; Salinas and Salazar 2006; Salinas et al. 2006).

The goal of this article is to present and discuss part of our analytical model and the main results of our ongoing research on the prehistoric mining systems of San José del Abra and its transformations through time. Due to space limitations, we will focus on the technological organization of the mining operations we have studied, including the identification of activity areas and the functional analysis of its lithic instruments (hammerstones). The archaeological evidence discussed comes from the Cerro Turquesa area and includes the sites Cerro Turquesa or AB-83 of the Late Formative Period (ca. 100 AD – 1200 AD), AB-82 of the Late Intermediate period (ca. 1200 – 1450 AD) and AB-22/39 and AB-37 from the Late Period (ca. 1450 – 1540 AD).

Study area

San José del Abra is the name given since the 19th Century to an area of regional relevance in terms of its mining operations, from pre-Hispanic times to the present day (Figure 1). The locality is characterized by a rocky massif running in a North-South direction, known physiographically as Sierra del Medio, with an average elevation of 4,000 meters above sea level. This range is cut by many transverse *quebradas*, and their spreads form alluvial planes to the East and West of the massif.

The dominant climate is that of a High-Altitude Marginal

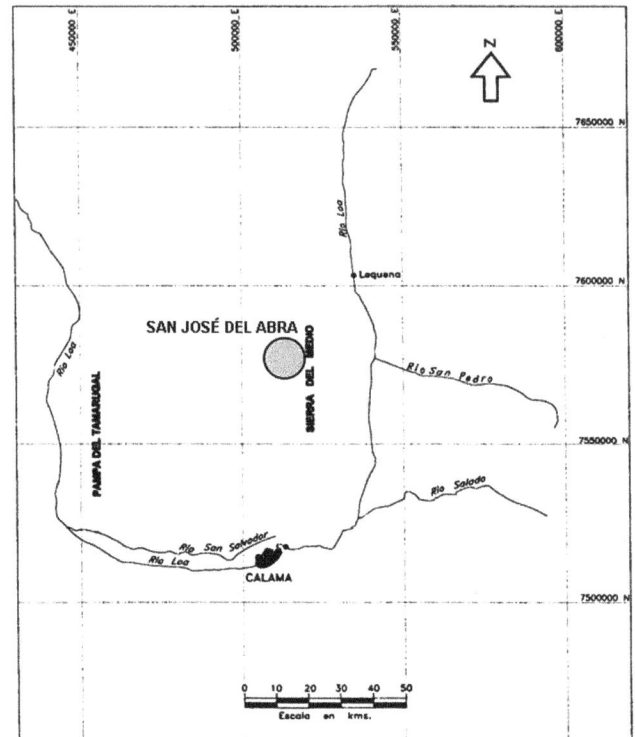

Fig. 1: The Atacamenian north of Chile and the study area, San José del Abra.

Desert, though higher zones may be classified as High-Altitude Steppes. This means that the mean year-round temperatures are low, with an average of 10.1 °C and a mean relative humidity of 27%. Daily temperature oscillations are large in amplitude, while year-to-year precipitation is highly variable, though always moderate. Precipitation occurs during two periods: a summer rainy season, generally between January and March; and a winter snow season, between May and July.

Given its climatic and orographic conditions, the study area presently exhibits little vegetation other than that found at the bottom of *quebradas*, where there is greater availability of water from underground sources and some springs. Vegetation is mostly composed of xerophyte shrubs between 30 and 70 cm in height, as well as short plants and cacti. Considering the high altitude of the locality, the low winter temperatures and the dominant type of soils –generally rocky and immature–, it is evident that San José del Abra has little agricultural potential, and presents difficulties for permanent settlement. It is because of this that, throughout its history, this area has been rather marginal within Atacamenian territory, having been occupied fundamentally with the objective of obtaining certain specific resources, such as hunting and, of course, copper mining.

From the geological point of view, San José del Abra presents volcanic and sedimentary rocks from the Jurassic period, intruded by various intrusive events, providing them with diorites, granodiorites, monzodiorites, monzonites and breccias present today (Ambrus 1977; Moraga 2000; Ulricksen 1990). The main mineral deposit present in this area is a porphyry copper composed of a significant layer of copper oxides on top of an even larger deposit of sulfides. The oxide layer, focus of pre-Hispanic mining operations, presents a wide variety of minerals, predominantly chrysocolla (SiO_3 Cu $2H_2O$), brochantite ($4CuO$ SO_3 $3H_2O$), pseudo-malaquite ($6CuO$ P_2O_3 H_2O) and turquoise (CuO $3(Al_2O_3)$ $2(P_2O_5)$ $9(H_2O)$), among others (Moraga 2000).

Cuprite (Cu_2O) and native copper ($Cu0$) have also been detected in certain areas, though mainly towards the supergene enrichment layer (Gerwe et al. 2003). The mineralization can occur in the mineral veins that control the oxidation zone (approximately 10% of the deposit), in clays and iron oxides with low-grade copper ore (10%) or in an oblong ore body, running in a North-East direction, that contains the main deposit (80%) (Gerwe et al. 2003). Finally, there is also an exotic area which formed as a result of the partial erosion of the main bulk of the deposit.

In the present paper we will focus on the evidence for mining activities found in the area known today as Cerro Turquesa, since these sites have been most systematically studied to date. (Figure 2). This area is a large, flat expanse to the West of the main mineralized body, located between the Ichuno and Casicsa *quebradas*. Most common are monzodioritic rocks with a high content of biotite and potassic feldspars, as well as smaller loci of breccias, granites, granodiorites and aplites. The surface mineralization specific to this area consists of exotic copper oxides transported from nearby hills (mainly chrysocolla and brochantite). Nonetheless, in addition to these minerals, there is a significant presence of turquoise in the original oxidation zone. It is also important to bear in mind that, in this sector, intrusive rocks have been severely altered by high temperatures, a result of a localized hydrothermal quartz-sericite intrusion. This event generated clays that trapped copper ore present in hydrothermal solutions, and thus there are no important veins in this area, but rather a diseminated zone of mineralization. Another consequence of this hydrothermal intrusion is that the hosting rock in two of the three prehispanic mines of Cerro Turquesa is a soft and crumbly substrate, milky-white in color, with occasional reddish-brown tones due to the presence of limonites (Ox-Fe) hosted in geological fractures.

Synthesis of local prehistory

The North of Chile is one of the richest regions in the world in copper ores. The exploitation of this deposits is today one of the main props of the country's economy. But these ores have been mined for centuries as part of particular socioeconomic regional contexts, and using different technological devices.

Archaeological research has demonstrated that in the

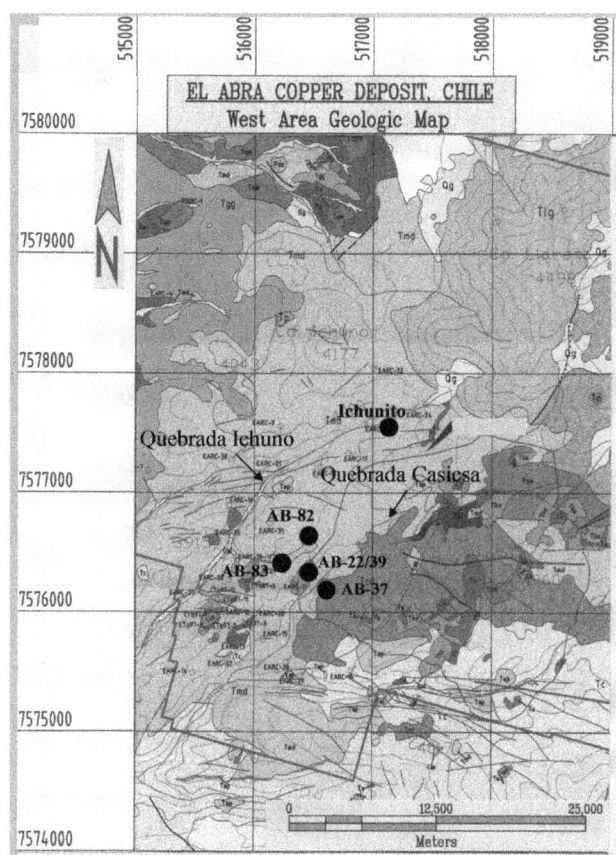

Fig. 2: Geological map of San Jose del Abra and the sites discussed in the text.

Atacama Desert copper ores were mined for the first time during the Late Archaic Period (ca. 3.500-1.500 B.C.). During this period, Hunter-gatherer communities were experimenting significant social, economic and cultural transformations as a result of the development of technologies of animal domestication and increased sedentarism (Núñez 2006; Soto 2006). There is yet no direct evidence of mining activities for this period, but archaeological finds demonstrate that copper ores were in use mainly for the local elaboration of beads and pendants, that is, as semiprecious stones. It is interesting to consider the possibility that copper ores are playing a role as signifiers for social differences that occur as a result of the abovementioned transformations. Beads and pendants also take part in long-distance trade institutions that appear in the Andes with camelid domestication. The fact is that throughout prehistory, copper minerals and metal will be used mainly as symbols of social and political status and as exchanged prestige goods.

Metallurgy itself will appear in Northern Chile only during the Formative Period (ca. 1500 B.C. – 500 A.D.), together with other new technologies such as textiles and pottery, as well as horticultural practices. Metallic objects are made of copper or gold and are few in number. They are limited mainly to ornaments and ritual objects, which have been found in some exclusive burials within cemeteries. Formative period demand for copper ores increased in regard to Late Archaic times, not only due to the development of metallurgy, but also since the use of copper ores as semiprecious stones also witnessed a notorious increase. In fact, some sites are producing massive quantities of beads both for long-distance exchange and local consumption (Rees 1999, Núñez 2006; Núñez et al. 2006).

In the nuclear areas of the Atacamenian territory, the Middle Period is characterized by the social and cultural influence of the Tiwanaku state of the Bolivian altiplano (Berenguer and Dauelsberg 1989). Metal objects of gold, copper and silver (and different alloys including a tertiary bronze) can be found with more regularity in the cemeteries, but always associated with some of the deceased (Lechtman and McFarlane XXX). Beads and pendants of copper minerals are also common, as well as their use as incrustations in wooden and textile artifacts. Tiwanaku's influence was not homogeneous throughout atacamenian territory. In more marginal areas, such as San Jose del Abra, Tiwanaku's influence is virtually invisible, and local families maintained a traditional Formative way of life and technologies until the Late Intermediate Period (ca. 900 – 1.450 A.D.). Nevertheless, they were not isolated from regional exchange networks, and some of these small communities produce surplus copper ores that move throughout the territory in camelid caravan trails.

During the Late Intermediate Period the amount of metal objects seems to decrease (Schiappacasse et al. 1989), but this is only apparent since even though objects produced in other regions are more scarce, metallurgy is still represented in local sites and metal objects, mainly of copper appear in almost every cemetery. Beads and pendants also continue to be produced, and the use of crushed copper minerals in local rituals is very popular during this period. Besides this, there is now evidence of the use of copper ores as pigments in regional rock art (Sepúlveda and Laval 2010). Therefore, it is likely that mining activities increased during this period as a result of a more diverse demand and use of cooper ores. In fact, local populations maintain mining colonies in far distant areas, such as El Salvador, nearly 500 km to the south of the core Atacamenian zone (Westfall and González 2006).

The inka empire, known as Tawantinusyu, dominated Atacamenian region at least from 1450 AD until the arrival of the Spaniards in 1536 AD. It is well known how the incas increased mineral and metal production throughout the empire in order to produce goods that would be exchanged as part of reciprocity relationships with local leaders and thus gain their loyalty (Earle 1994). Once again then, we see an increase in local production.

In short, the demand for copper ores increased constantly in the atacamenian region throughout prehistoric times. Even though new uses wer incorporated in the different periods, what seems to have remained quite constant is the fact that both mineral and metal objects were mainly used as symbols of identity/status and in ritual settings. Thus, mining and metallurgy were very much associated with increased social differenciation, political leadership and ritual life, unlike the Old World where one of its main uses were always domestic tools, warfare and transport industries (Lechtman 1991, González 2004).

The sites in their regional socio-historical context

Cerro Turquesa (Site AB-83) (Figure 3)

The site consists of a nearly nine-meter vertical shaft associated with two archaeological structures of little height, where domestic and productive activities were concentrated. The site also presents an associated area of waste rock in the vicinity of the main shaft, though it is most significant downhill from this point, where a flat-surfaced terrace and retain wall were built. Analysis suggests that this extraction operation mainly produced turquoise and copper oxides, such as brochantite and chrysocolla. A total of 23 stone instruments, generically termed hammers, were recovered at this site. Almost all of them were obtained from superficial collections, except for a few recovered during excavations.

The six available radiocarbon dates place the exploitation of this mine between 200 and 1200 AD approximately, whereas the peak of the mining activities can be located between 800 and 1200 AD (Salazar et al. 2006), during the *Carrazana* phase of local prehistory, a transitional period between the local Late Formative and Late Intermediate Periods (Berenguer 2004). Considering this period's settlement pattern in San José del Abra, as well as the materiality associated with other sites from the same period, we have previously proposed that during Carrazana Phase the area was inhabited by a few *atacamenian* families, that would remain in San José del Abra a few months during the year, dedicated to complementary economic activities such as hunting of camelids, herding and the extraction and processing of copper ores (Salazar et al. 2006). The seasonal occupation is demonstrated by the size of the site, the organization and density of the domestic deposits, the degree of sedimentation in the formation of the archaeological deposits and the variability of pottery types that include all those at use in the main occupation sites of the region. Towards 1100 AD some metallurgic activities are evident in the area, concentrated at the Ichunito site. These, however, did not develop significantly, and the basic pattern is that the productive chain ended after a secondary crushing and sorting of the ore. Thus, there is little doubt that the selected material was mainly transported out of the study area, although we have no regional evidence concerning the location or details of subsequent steps in this production process.

Site AB-82 (Figure 4)

Located on the same Cerro Turquesa esplanade, and some

Fig. 3: The Cerro Turquesa (AB-83) site from the Late Formative Period (ca. 200 – 900 AD). Note the open pit and the structures in the back part of the Picture.

300 meters to the North-East of the previously discussed site, AB-82 is a mining operation consisting of four shallow open pits, all located sequentially downhill in a SE-NW (112°) direction. Apart from the aforementioned openings, each with its associated deposit of waste rock and tailings, there are two crushing areas, one domestic structure and two small circular structures of unknown function.

The total area of the site is approximately 1500 m², and the main product of the mining operation was turquoise. At AB-82 a total of 35 hammerstones were recovered, almost all of them from surface collections, save a few examples recovered from excavation.

One radiocarbon date is currently available for AB-82, corresponding to a calibrated result of 1290 to 1420 AD (Salazar 2002-2005), and this places the exploitation of the site in the local Late Intermediate Period. Other investigations on a local and regional level indicate that during this period San José del Abra was inhabited less permanently than during previous centuries, possibly also by a few *atacamenian* families coming from small hamlets and *estancias* of the upper Río Loa. Though it is indeed possible that herding-related activities were carried out in this period, and there is limited evidence supporting hunting

activities as well, the central motivation that drove this people to access San José del Abra during Late Intermediate times was the exploitation and processing of mineral ores. Metallurgical activities were also carried out in the of Ichuno and Agua de Llareta quebradas, even though there is no evidence for metalworking in these areas.

San José del Abra Mining Complex (Sites AB-22/39 and AB-37) (Figures 5 and 6)

Sites AB-22/39 and AB-37 were discovered originally by Lautaro Núñez and they are part of the San José del Abra Mining Complex of Inca times (Núñez 1999; Salazar 2002 and 2002-2005). Site AB-22/39 is a large indigenous mine, defined by four related elements in an area of more than 3000 m²: mines or extraction areas (16 cuts, 5 shafts and a few possible galleries); waste rock and tailings associated with each extraction operation; 8 retain walls (for rainwater, colluvium, and waste rock from previous operations); and 8 working platforms known locally as *canchas*. The main mineral extracted at the site was turquoise and, to a lesser extent, chrysocolla. At site AB-22/39 a total of 359 hammerstones were analyzed, all recovered from surface collections.

Fig. 4: Site AB-82, of the Late Intermediate Period (ca. 900 – 1450 AD)

Site AB-37, on the other hand, consists of an architectural ensemble composed of nine stone-walled structures, built on artificial terraces on the South-East slope of the Casicsa quebrada. Four of these structures were used for domestic activities (preparation and/or consumption of food), while the rest are areas dedicated to ore crushing or the dumping of gangue separated from the rocks extracted from the nearby indigenous mine AB-22/39 (directly opposite AB-37). Geological analysis of these rocks shows that they come from the main shafts of site AB-22/39, and from this we conclude that the construction of site AB-37 is contemporary to the peak of productive activities at the indigenous mine. An available radiocarbon date for the site provided a calibrated interval of 1420 to 1660 AD. The presence of local Inca and altiplanic ceramics at the site, as well as at the nearby mining camp (Inkawasi-Abra), thus prove that the largest extraction operations occurred during the Incan domination of the region (Núñez 1999; Salazar 2002).

Local and regional data suggest that during the Late Period, the volume of production at San José del Abra was notably increased, with mining operations concentrated at sites AB-22/39 and AB-99. Given that the organization and logistics of operations were in the hands of the Inca state, a larger population was able to gather at San José del Abra, dedicated exclusively to mining, and supplied by the state with the surplus collected from other localities in the *atacamenian* region.

Analytical framework

In previous papers we have proposed that in order to understand the organization of productive mining systems and their transformations through time it is necessary to consider the interaction of four sets of fundamental variables, namely, the natural and geological environment, the technological and technical system, the socioeconomic organization of the group and its cultural dimension (Salazar 2003-2004; Salazar and Salinas 2007). In this article we will center on the analysis of two specific variables within the technical and technological aspects of a mining system: the functional variability of its instruments (in this case lithic artifacts known generically as hammerstones) and the spatial organization of activity areas. In order to archaeologically approach these variables we defined material attributes that were recorder at each of the aforementioned sites, and which have served as a base for our comparative analysis of the three prehistoric periods identified at our study area. In what follows, we will discuss these variables and their archaeological expression.

Fig. 5: General view of the AB-22/39 mine of the Late Period (1450 – 1536 AD).

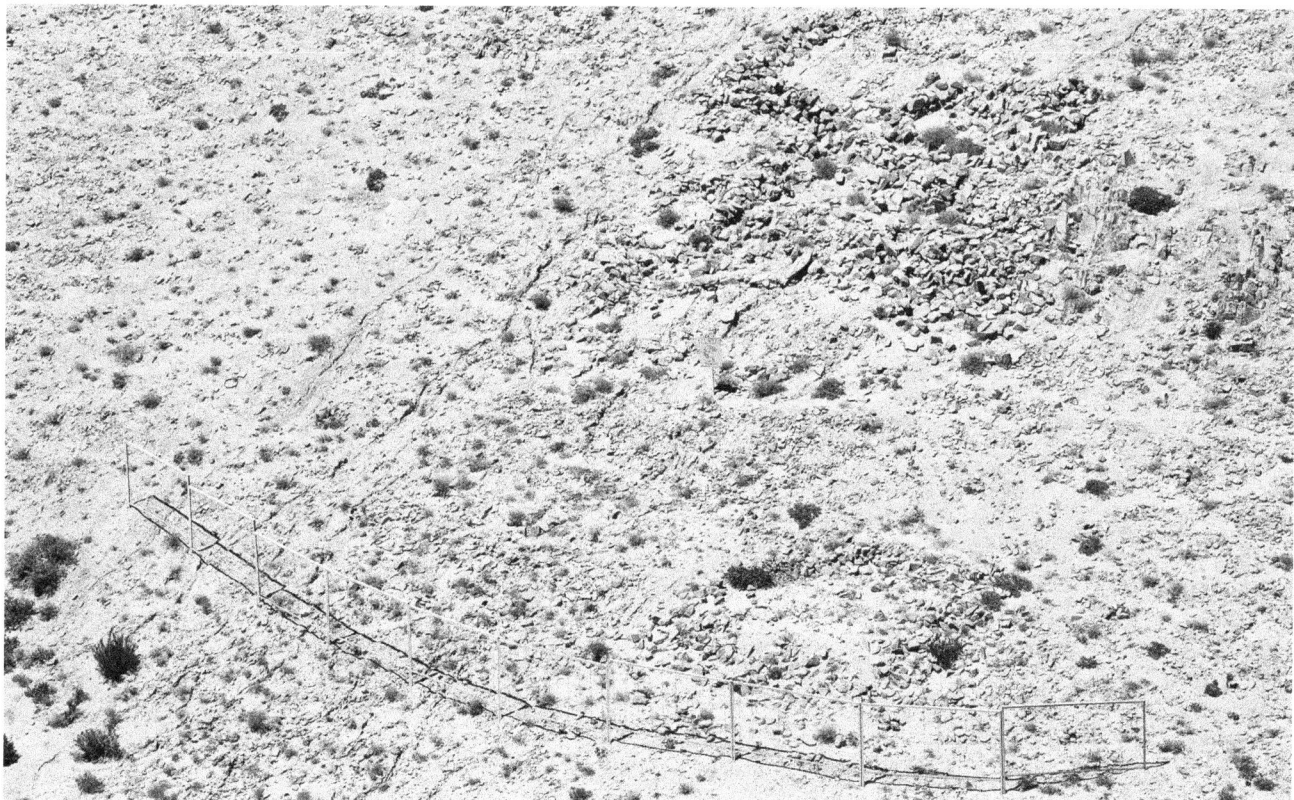

Fig. 6: General view of site AB-37, crushing areas from the Late Period.

Variable 1: Towards the functional analysis of hammerstones

Due to the lack of functional information provided by the typological perspective generally used to describe hammerstones from prehistoric mining (Craddock 1990; Núñez 1999; Esperou 1992; Timberlake 1990; Pickin 1990; Thorburn 1990) it was necessary to find a method that would allow us to lay the foundation for a reliable interpretation of the activities and processes of which these hammers were a part in the mining operations of the past (Salinas et al. 2006; Salinas 2007).

That is to say, we have been interested in determining how and in what way these tools were used in the context of a pre-Hispanic mining operation. The answers to these questions clearly concern the instruments' functionality, and thus it is relevant to identify archaeological evidence of the past use of these instruments. It has been stated before that the hammerstones present at the sites we have studied are functionally diverse (Núñez 1999; Salazar and Salinas 2005), but the question is to what functional categories do they belong? What parameters should be used to estimate this diversity? And, finally, how are these elements applied to the study of pre-Hispanic mining technology and the transformations it underwent throughout its history?

In order to analyze their functionality, we consider that the organization and internal variation of artifactual sets is a result of factors directly related to the performance of each instrument and its association to certain specific tasks within the production process (Salinas 2007). Therefore, our observations have concentrated so far on three fundamental variables or attributes, that we consider of key importance in order to determine this Functional Aptitude of the hammerstones. The variables we have used in our analysis are the morphology of the active surface, the raw material or type of rock and the total dimension of the artifact. We consider that these three variables give us a plausible initial idea of how suited a given hammerstone is for a specific role in the mining production process. Even though we should also consider the weight of the hammers and their hafting system as variables relevant to the mass and potential impact force of these instruments, so far they have not been included in our analyses (in the second case, given their absence from the archaeological record).

On the other hand, the determination of an instrument's function is not an intrinsic property of the artifact, but rather of the activities it has been used in. Hence, investigating its function requires us to be able to correlate the artifacts with certain actions and the materials they have been used on. Therefore the reconstruction of the Functional Aptitude of an artifact through the analysis of the three aforementioned variables, allows us to formulate hypotheses about the specific functionality of the instruments that have to be supported empirically with use wear analysis.

a. Morphology of the Active Surface (AS). Functionally, the

shape of the AS serves two main purposes. In the first place, it allows control over the impact area. More importantly in terms of the mechanical modeling of the action of this class of artifacts, the shape of the AS determines the precise contact area between the instrument and the rock. This contact area is reduced or increased depending on the type of AS used, and consequently the total force of the impact will either be concentrated or dissipated (a wide AS area will result in a less localized impact force, and vice versa).

b. Raw Material. This variable is extremely relevant, especially since it allows the determination of whether or not the user has an empirical knowledge of the physical properties of the lithic raw materials used to create a hammer. In our case, through mechanical tests similar to the Charpy test (Hertzberg 1976), we have recognized differences in the properties of the rocks used as hammerstones (Salinas et al. 2006). These results suggest that the selection of two groups of lithic raw materials (Silicificated Andesites and Granodiorites) was mainly due to the criteria of mechanical resistance, which is essential for an effective fracturing of the host rock.

c. Dimensions. The size of a hammer is directly related to its mass and, therefore, to the impact force. It also influences indirectly the general morphology of the artifact, as well as its active edge. A hammer's movement follows a trajectory that is ideally circular and the force it is able to apply is determined by the length of this trajectory, its speed and the total mass of the artifact. If the speed is limited to what a human operator is normally able to generate, it is clear that the variation in the size (and mass) of the hammerstone will be an adequate strategy for the increase or decrease of the applied force[1].

d. Use-Wear Analysis. In order to understand the functionality of a pre-Hispanic mining instrument we must be able to identify its relation to a given action, as specifically as possible. This implies the ability to associate instruments to one of the sequential phases of a mining operation (extraction, primary crushing, secondary crushing and so on), and for these inferences we must turn to experimentation. Using raw materials identical to those found at the archaeological sites, we have carried out preliminary experiments for the three aforementioned phases, considering in our experimental design the control of variables such as shape, direction of trajectory, type of host rock and time of use. The use wear marks observed were Abrasion, polish, Scars and Splintering, and these were evaluated in terms of morphological attributes, size, distribution and quantity. Our observations were macroscopic only, due to the impossibility of introducing the archaeological samples into conventional optical instruments. It is important to consider that the results of our experiments are still preliminary, and their application

---

[1] Another relevant variable is the length of the hammer's shaft, since the longer it is, the longer the trajectory followed by the head and, therefore, stronger the impact force will be. However, as we have already stated, this variable has been excluded from our analysis since we have no archaeological evidence of the hafting system used for the stone hammers.

is, for now, limited, until a greater number of experiments can be carried out and compared to a more representative archaeological sample.

In summary, we have worked with four specific attributes for determining the functional variability of hammerstones; the first three are related to the material and morphological properties of the artifacts, while the fourth is based on the principles of use-wear analysis. We suggest that the combination of the first three of these corresponds to the functional aptitude of an instrument and might therefore be correlated with a specific use in the systemic context. We may assume the existence of specialization in the archaeological record of mining activities when we observe a greater selectivity applied to any of these variables –or all three at once– (Salinas and Salazar 2008).

Variable 2: Organization of Activity Areas

We will work under the assumption that each step in the mining operative chain is carried out within previously-defined spatial limits; this means that different productive tasks are only carried out in a specific sector of the total space of the settlement, even when these may overlap (Salazar et al. 2005; Salinas and Salazar 2008). We have therefore set ourselves the task of identifying the specific productive activities carried out at each site and their spatial organization and layout. For this we make use of the concept of activity area, which we have recognized archaeologically through the analysis of four fundamental variables and their vertical and horizontal frequency and distribution:

Mineral debris: This corresponds to all derivatives of the extraction, reduction and selection of the ore, either waste rock or gangue. The different phases of the production process (extraction, primary crushing, secondary crushing, etc.) will generate different types of debris in terms of their frequency, size and shape, and their mineral content or grade. Though the analyses of these variables must take into consideration local geological conditions of the mining operation, as well as ore processing technologies, in general terms the initial phases of the productive process will generate a larger amount of waste, in larger pieces and in more irregular shapes than later phases.

Artifactual debris: Artifactual debris refers primarily to broken lithic artefacts and their associated stone flakes, whether from their manufacture or their use. Based on ongoing experimentation, we have attempted to differentiate both kinds of debris based on specific attributes, such as the size and shape of flakes, as well as the presence of cortex and attributes such as size and type of talus (Salinas 2007).

Instruments: After the functional analysis performed according to the attributes described above (variable 1), we have tried to identify the spatial distribution and associations of each functional category.

Structural features: We refer to the presence in the archaeological record of man-made constructions or alterations in the environment due to the mining activities. Typical examples are the working platforms usually associated with the secondary crushing of the ore (known in Spanish as canchas), structures, mines, tailings and other general features.

As we can see, we study the activity areas of ancient mining operations by means of the relationship between variables traditionally used to this effect (distribution and frequency of artifacts, ecofacts and features), but applied to the specific case of small-scale mining.

Results

We now present the main results of our investigation, organized according to cultural period for greater clarity.

Late Formative Period

As a result of the analysis of the raw materials used in the hammerstones of the Cerro Turquesa site, we concluded that there are two main lithological groups that encompass almost the entire sample. On the one hand, there are rocks from the granodiorite family, available near the site and, on the other hand, a group of silicificated andesites from the San Pedro de Conchi ravine, located about 12 km from the sites studied in this article. These data indicate that knowledge of the physical properties of the rocks that were used had been mastered by the Late Formative Period and was put to use in the technological decisions of the miners. The mere presence of silicificated andesites, which are alochtone rocks of good quality and reliability, is an undeniable indicator of this, demonstrating the existence of selectivity criteria based on empirical knowledge of lithic material. Nonetheless, it should be kept in mind that, for this site, the frequency of silicificated andesites is lower than that of local rocks from the granodiorite family (Figure 7).

The aforementioned data suggest that populations from the Late Formative Period could have had limited access

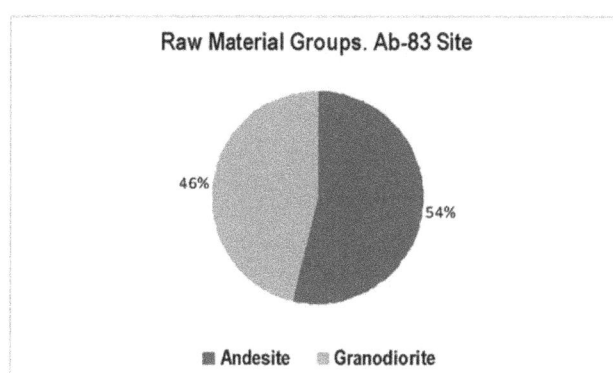

Raw Material Groups. Ab-83 Site

46% 54%

■ Andesite ■ Granodiorite

Fig. 7: Distribution of hammerstones' raw materials from the Cerro Turquesa site (AB-83). G: Granodorites, A: Silicificated Andesites.

to higher quality material (silicificated andesites), thus complementing it with lithic material available near the site (granodiorites). The andesites, on the other hand, were probably used to fulfill more specific requirements of the mining operation but do not seem to have developed significant properties of functional specialization, at least at this level.

With regards to the size of the hammerstones, the instruments from Cerro Turquesa are, in general, of relatively large dimensions, especially along their longest axis (X = 13.56 cm). Concerning their width and thickness, the average width is within the range observed at other sites (X = 6.18 cm). The thickness, on the other hand, is greater on average than that observed at all other sites (X = 7.91 cm) (Table 1).

The analysis of the morphology of the AS from the Late Formative Period revealed limited variability. Almost the entire sample falls within three of the five categories defines: abrupt (an edge angle between 60° and 90°), acute (an edge angle less than 60°) and flat (an edge angle of 90°). This scarce variation may have been partly conditioned by the nature of the mining operation and the qualities of the host rock, which has been highly altered by local hydrothermal processes. Despite this, it could also be interpreted as a situation derived from the restricted access to andesites and the scarce specialization of mining activities during the period. In this sense, data indicates that nodule selection was guided by the search for AS that were well adapted to multifunctionality.

The previous point is partially corroborated when one considers the scarce functional specificity suggested by the analysis of use wear for this period. This can be explained as a result of the instruments' multifunctional use. Indeed, we consider that the formation of diagnostic traces on the artifacts we analyzed could have been hindered, since these were used for several different tasks throughout their useful life.

In terms of productive activities and spatial organization, our studies indicate that the Cerro Turquesa site was the location of turquoise and brochantite extraction, as well as primary, secondary and, possibly, tertiary crushing. This is supported by the lesser size of discarded debris (gangue) and its increasing homogeneity in specific areas within the site. Considering that the dates available for the Cerro Turquesa site indicate approximately 1000 years of working, stratigraphy suggests that during the height of the mining operation (800-1200 AD), primary and secondary crushing was carried out near the mine entrance, and waste rock was discarded along with domestic debris (charcoal, ceramics, animal bones and vegetables, among others). Thus, though a certain spatial segregation of productive activities is apparent, these are not delimited by walls or equivalent structural features but, on the contrary, are carried out at no more than a few meters from areas destined to entirely domestic activities, such as the preparation

AB-83	Lenght	Thickness	Width
Average Size	13,569	6,185	7,923
sd	2,937	1,781	2,058
Andesite	12,933	5,517	6,75
Granodiorite	14,114	6,757	8,929

Table 1: Table showing average sizes for andesite and granodiorite hammerstones from the AB-83 prehistoric mine.

and consumption of liquid and solid foods (Salazar et al. 2006). It is important to consider that there is no evidence for activities corresponding to later phases of the mining production chain at Cerro Turquesa, so it is not possible to determine if the purpose of this operation was to supply the metallurgic or lapidary industries, or both.

Late Intermediate Period

The lithic raw material situation during the occupation corresponding to the Late Intermediate Period (Site AB-82) shows continuity with respect to the previous period, in terms of the predominance of the two main rock families (granodiorites and silicificated andesites). Despite this, during the Late Intermediate Period, the use of silicificated andesites increases to cover all functions derived from the requirements of the mining operation (Figure 8). This greater presence of andesites is probably a result of a change in the logistic organization of mining activities during this period, a point we will come back to further on.

With regards to hammerstone sizes, though there is a certain heterogeneity evident in the set, average sizes are generally relatively low (Average length= 12; Average Width= 10 cm; Average thickness= 6.89 cm), especially if they are compared to the dimensions observed at the Cerro Turquesa site, a previous occupation from the Late Formative Period (Table 2). The reduction in hammerstone size is related to another interesting phenomenon, mentioned previously, namely the increased proportion of andesite raw materials

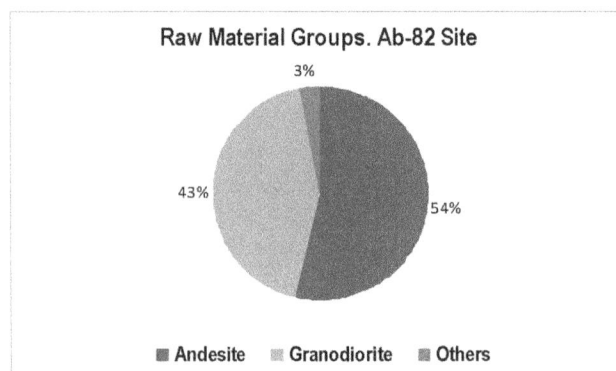

Fig. 8: Distribution of hammerstones' raw materials from site AB-82. G: Granodorites, A: Silicificated Andesites, O: Other raw materials.

AB-82	Lenght	Thickness	Width
Average Size	12,106	6,014	6,894
sd	3,187	1,044	1,960
Andesite	12,721	5,932	7,042
Granodiorite	11,267	6,107	6,627

Table 2: Table showing average sizes for andesite and granodiorite hammerstones from the AB-82 prehistoric mine.

at the site during this period. This situation is more or less the inverse of that seen at Cerro Turquesa.

The thickness of hammerstones from AB-82 is within the range of average variability observed in the sets from all other studied sites. This fact suggests that selection of the thickness of nodules for the creation of hammers is largely determined by the hafting system, since it remains more or less constant from one settlement to another in the area.

It should be noted, however, that if we consider the dimensions of stone hammers (length and maximum thickness) for each group of raw materials, we find that the silicificated andesite hammers at AB-82 are more robust and larger by a centimeter, on average, than those made from granodiorite Table 2).

Though the difference may not seem important, in terms of volume, the increased mass and the differences between the two materials may be functionally important. This variation can be considered as a preliminary expression of functional specialization of the mining operations of the Late Intermediate Period, and two interpretations have been put forth.

The first interpretation concerns the need to take advantage of the properties of andesites for specific functions. We have considered this an incipient specialization of the artifacts, though not necessarily of the mining operation's layout. In a certain sense this has to do with a state of experimentation with the uses and advantages that might be derived from the physical properties of the chosen material. Though we assume that these properties have been known since the Late Formative Period, what we are advancing as a possibility is that the conditions for the development of a process of functional specialization were improving, whereas this had not been the case previously, at least not according to the available archaeological record.

The second interpretation takes into account the fact that different demands are imposed on the workers by the mining operation itself, including harder host rock than that at Cerro Turquesa, along with an improved supply logistics by the local population. This last fact may have provided the opportunity for selecting nodules with specific attributes that were required to face the specific geological conditions of the mine. In this sense, we suggest that the nodules are collected with a previous idea of what constitutes an

adequate set of hammerstones for the mining operation requirements.

Both interpretations indicate an important differences between the way in which mining at San José del Abra is carried out between the Late Formative Period and the Late Intermediate Period. This change is related to the implementation of new strategies of technological organization and, in this specific case, the way in which the nodules that will be used in the manufacture of stone hammers are selected. Given that initial functional specialization occurred at AB-82, the large variation in size can be understood as a result of the aforementioned experimentation. In a certain sense there is an exploration, as far as conditions allow, of the potential of known materials, in terms of the variation of their physical attributes, such as the size of the hammerstones.

During the Late Intermediate Period occupation of site AB-82, there is also a slight increase in the variety of AS, since four out of the five defined categories are present. In a similar way to what has been seen at the Cerro Turquesa site, the acute and abrupt AS morphologies prevail. Artifacts with a flat AS morphology appear infrequently in the record and hammers of convex AS are practically non-existent. This variation is especially apparent in artifacts of silicificated andesite, in such a way that we believe it fits the pattern of an incipient functional specialization, facilitated by improved access to raw materials.

Despite this, preliminary analysis of use wear indicates that the choice of AS did not take into account specific phases of the production chain. This is a result of the fact that most hammerstones at site AB-82 again showed practically no diagnostic use-wear traces, probably due to multifunctionality, as already proposed for the Cerro Turquesa site. However, seven andesite hammerstones could be classified as belonging to the phases of extraction, primary and secondary crushing. This is basically deduced from the large size (over 20 mm), the distribution (1-2 flaking scars per artifact) and the rectangular morphology of the use-wear traces observed, which were contrasted with experimental results.

The use wear marks identified on hammers at site AB-82 –though few– are consistent with analysis of the mine's debris, since in this case, their size and distribution indicates that they were the outcome of extraction, primary and secondary crushing activities. The productive chain is interrupted at this stage and later phases –whether they be metallurgic or lapidary in nature– must have been carried out at other localities, as it was in the Cerro Turquesa site. Unlike the previous period, however, at site AB-82 secondary crushing is segregated from the extraction and primary crushing areas, and is even delimited by low walls fashioned as wind breaks, found a few meters from the pit entrances. This fact once again suggests a possible incipient functional specialization at this site, since the productive chain is more spatially segregated than during the earlier

Raw Material Groups. Ab-39 Site

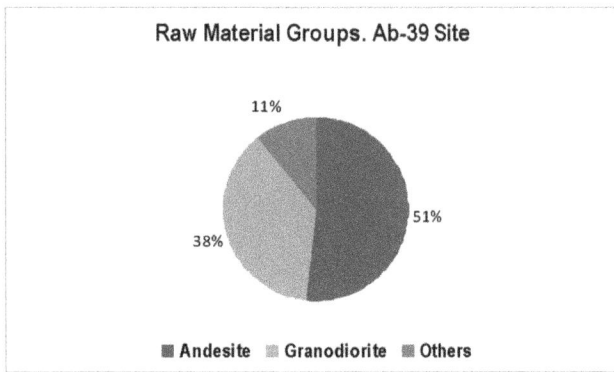

Fig. 9: Distribution of hammerstones' raw materials from site AB-22/39. G: Granodorites, A: Silicificated Andesites, O: Other raw materials.

period. Furthermore, domestic activities during the Late Intermediate Period are not carried out in areas destined for mining activities, as it was the case during the Late Formative.

Late Period

In terms of the raw materials used for the mining tools, during this period both principal groups are still in use, but in this case there is a clear preference for silicificated andesite and for other non-local raw materials, and this has logistic and technical implications (supply and transport) (Figure 9). It also reinforces our statement regarding the selection of rock type based on their relative physical advantages (Salinas 2007; Salinas et al. 2006).

The size range of the hammerstones from sites AB-22/39 is more varied than that from previous occupations, evidently as a result of the greater sample size. Apart from this, all measured dimensions (length, width and thickness) presented larger sizes with respect to previous periods (Tables 3, 4 and 5).

On the other hand, during Late Period the full variety of active surfaces is evident, a sign of the complex way in which the operation was organized. This lends support to our hypothesis of specialization enabled by the improved access and transportation of raw materials during this period. An adequate flow of nodules to the sites where they are needed allows for the selection and distribution of hammers with AS suited to the requirements of the mining operation. Furthermore, it may be the case that instead of the versatility observed as a technological strategy during the Late Formative Period, there is a preference for efficiency and specialization during the Late Period operations.

The correlation between the aforementioned attributes indicates that all the variability in AS can be found in each size range, even though there is some degree of correlation between small hammers and flat active edges. The virtual absence of flat edges on large hammerstones is supplementary evidence for our interpretation of functional

specialization in this set of artifacts. In this sense, it is necessary not only to identify the hammerstones, but also the substrate on which they were used. Use-wear traces have been useful to further this objective, though results are still not entirely conclusive (Salinas 2007).

Indeed, the presence of a certain diversity of functions during the Late Period is supported by the analysis of use-wear traces, since there are a greater number of instruments with diagnostic use-wear marks. Artifacts with superimposed wear marks and initial stages of surface polishing, similar to that of multifunctional instruments from previous occupations (Cerro Turquesa), were also identified.

In terms of production activities, evidence from the Late Period again shows continuity with respect to previous periods. In the first place, sites AB-22/39 and AB-37 were destined for extraction, primary crushing, secondary crushing and selection of colored ores, at which point the operational chain ended.

The correspondence between activity areas and the function attributed to different stone hammers as derived from dimensions and use-wear analysis is generally consistent. Thus, for example, areas associated with the first phases of the mining operation (extraction and, in one case, primary crushing) present the largest average sizes recorded for all measured dimensions. Considering that the primary phases of extraction require a large amount of work to fracture the host rock, it would seem that size is the deciding factor when it comes to increasing the hammers' fracturing capacity (see Tables 3, 4 and 5). Furthermore, use-wear analysis also indicated that in this areas we find hammers splintered according to initial phases of the productive process (Large, rectangular, non-overlapping, non-adjacent negatives on the active surfaces).

It is interesting to note that during the Late Period there is a significant increase in the magnitude of extractive operations and their associated activities, and that at the same time the productive activities are spatially segregated, in such a way that areas of extraction are more clearly separated from crushing areas, and these, in turn, from areas of ore collection. Site AB-37, for example, which was dedicated to secondary crushing of material extracted from AB-22/39, is about 50 meters away from the former, on the opposite slope of the Casicsa *quebrada*. This fact lends further support to the existence of functional specialization at the level of workers dedicated to different phases in the productive process, since it is unlikely that the same miners in charge of extraction and primary crushing would travel over 50 meters to carry out the secondary crushing.

Discussion

The archaeological data gathered at San José del Abra for the Late Formative, Late Intermediate and Late Periods prove that in atacamenian pre-Hispanic mining technology

Lenght

Sector	1	2	3	4	5	6	7	8
Average Size	17,553	16,232	16,826	16,618	15,627	16,22	16,8	17,321
sd	3,570	2,609	2,600	1,586	2,882	1,933	3,575	3,703
Andesite	17,575	16,462	16,633	15,7	15,525	16,408	16,342	15,086
Granodiorite	17,32	15,867	17,082	17,3	16,229	16,54	***	21,025

Table 3: Table showing lenght average sizes for andesite and granodiorite hammerstones from the different sectors of the AB-22/39 prehistoric mine.

Width

Sector	1	2	3	4	5	6	7	8
Average Size	6,476	5,672	6,619	6,418	5,846	6,296	5,886	5,7
sd	1,425	1,039	1,397	0,577	1,367	0,974	1,550	1,481
Andesite	6,775	5,992	7	6,433	6,125	5,992	5,608	5,229
Granodiorite	6,31	5,133	6,473	6,344	5,814	6,85	***	6,25

Table 4: Table showing width average sizes for andesite and granodiorite hammerstones from the different sectors of the AB-22/39 prehistoric mine.

Thickness

Sector	1	2	3	4	5	6	7	8
Average Size	9,688	9,556	12,819	10,294	9,127	9,836	9,671	9,907
sd	1,953	2,011	16,641	2,472	2,362	2,176	1,921	2,192
Andesite	8,5	9,938	9,583	8,817	8,688	9,146	9,342	9,371
Granodiorite	9,91	8,511	10,4	11,289	9,129	10,74	***	10,15

Table 5: Table showing thickness average sizes for andesite and granodiorite hammerstones from the different sectors of the AB-22/39 prehistoric mine.

there exist elements that are constant through time and others that underwent a significant transformation (Salazar 2002-2005; Salazar and Salinas 2007).

In terms of lines of continuity, we have the use of granodiorite and silicificated andesite as raw materials for the elaboration of hammerstones. Also we may include the general morphology of the lithic instruments and, possibly, their system of hafting system. Similarly, the knowledge of metal ores and the stages of the productive chain present at extraction sites also show continuity through time. In the case of lithic raw materials, the presence of non-local silicificated andesites in the archaeological record of all three sites, as well as the knowledge of the local mineralogy, indicate that there was not only knowledge of the *know-what*, but also of the *know-how* of mining operations. This indicated the existence of a certain technological tradition in the study area.

Despite this continuity of atacamenian mining knowledge in San José del Abra, the variables analyzed also show significant temporal changes that must be causally

explained. As mentioned earlier, the frequency of silicificated andesites increases in the Late Intermediate and especially in the Late Period. It is interesting to note that this transformation is coupled with changes in other variables studied. Indeed, it is evident that as the locale's chronological sequence progresses, and in particular during the later occupation, the dimensions of lithic hammers present a greater variability, especially those made from andesite, and the same is true of the AS used. The Late Intermediate Period shows a middle ground point in between the Late Formative and Late Period data. All of the above indicates that we have an increasing process of functional specialization in the pre-Hispanic mines of San José del Abra, since the material requirements of mining operations are met with a growing variety of attributes and combinations of attributes in the hammerstones, that are better adapted to perform specific functions in the operative chain. Indeed, during the Late Period, the functional variability attributed to the different lithic hammers is generally consistent with activity areas corresponding to the different phases of the operation (extraction, primary and secondary crushing).

If this is indeed so, then during the Late Period the existence of an efficient system of procurement and transport of hammerstones from their sources was indispensable –accompanied, naturally, by other items necessary for the manufacturing of full hammers. Not only was the transported material quantitatively significant, but also it was sufficiently varied to allow users to choose among a variety of alternatives, and select those which were best adapted to the particular requirements of the general operation (with its segregated areas of activity and phases), as well as the particular demands of each phase, including contingencies and unforeseen events that arose during the course of operations.

Mining exploitation during the Late Intermediate period would thus appear to be a transitional phase, with evidence for a growing specialization given by a greater variability of attributes and a greater segregation of activity areas, but one which lies within parameters similar to those of the Late Formative operation. During the Late Period, on the other hand, the increased capacity for raw material and supply access logistics finally enabled a fuller expression of the variability that had been constrained during the Late Intermediate Period by limited availability of material.

During the Late Period not only is there the possibility of strategically implementing the operation, of anticipating the lithic technology that will be used, but also, there is the possibility of redesigning the action plan itself, making use of the abundance of functional options available. This varied stock of artifacts allows the resolution of ever more specific problems and situations, including the loss of functionally-specific instruments.

The process of functional specialization we have discussed is entirely compatible with the progressive increase in volume of mining operations during the three periods, as with the settlement patterns in the area. The latter is especially relevant, since it shows that while during the Late Formative period human populations carried out economic activities that were mixed in nature and complementary to mining (e.g. hunting and cattle raising), during the Late Intermediate period there is practically no archaeological evidence for these activities, which suggests that San José del Abra was accessed exclusively for mining (Salazar and Salinas 2007). On the other hand, during the Late Period, not only is mining activity dominant in the locality, but also the Incan state is in charge of financing operations, allowing, for the first time, the presence of miners exclusively dedicated to extraction and processing of mineral ores.

Considering the continuity evident in the mining know-how of the region, and the fact that artifactual and contextual evidence from the three periods leave no doubt that the populations in charge of mining operations were local, we posit that the Incan state made use of the consolidated atacamenian mining tradition, re-organizing it in order to articulate a new productive order, larger in scale and specialization, with the purpose of meeting the increased demands for copper ore that reciprocity imposed on the provincial administration (Salazar 2002).

Thus, our interpretation is that there is a local atacamenian mining tradition of great temporal depth, that changes over time as a result of the socioeconomic transformations of the region and the increasing importance of mining among the productive activities of the population that come to San José del Abra. Therefore, the differences observed in the designs of mining activities and their scales of production are not due to technological transformations, but rather sociopolitical and economic variables. These variables are those that allow new productive systems to develop in San José del Abra, from virtually identical geological and environmental settings and conservative technological traditions (Salazar and Salinas 2007).

Conclusions

By analyzing archaeologically a set of material variables considered relevant for the characterization of prehistoric mining, we have been able to further our understanding of the spatial and functional organization of the productive systems at San José del Abra, in Northern Chile, and its transformations between the Late Formative Period and the Late Period (ca. 200 – 1536 AD).

The data we have processed so far presents empirical support for the existence of an atacamenian mining tradition, with an important degree of continuity and conservatism in terms of the knowledge, techniques and technologies that are put into practice in the implementation of productive systems. However, we have also found that this was not a static and invariant tradition, but rather one that was modified and updated in relation to socioeconomic and political transformations that the region underwent during the last 10 centuries of pre-Hispanic history.

As we have pointed out in this article, the functional aptitude of stone hammers is one of the properties whose material manifestation changed throughout the area's temporal sequence. Indeed, during the Late Period we find a high variability in the range of hammer sizes, associated significantly with the variation of the AS and a considerable increase in the use of silicificated andesite as a raw material for their manufacture. It is quite likely that knowledge of mining conditions during this period guided selection criteria at the sources of non-local raw materials. Thus, the range of sizes and AS show a progressive transformation from a low selectivity during the Late Formative Period, through to the presence of several standard ranges for silicificated andesite during the Late Period.

We have interpreted these transformations of technological organization as evidence of an increasing functional specialization of mining activities, in such a way that miners selected attributes that effectively manage the material restrictions imposed by the lithological and minerological characteristics of their mines. We must bear in mind that during Formative times, the regional demand for copper ores was not as significant as during later periods, and especially that local families in San José del Abra were dedicated to diverse economic activities, including mining, hunting and herding. In later periods the inhabitants of San José del Abra were dedicated almost exclusively to the extraction and processing of copper ores.

One independent source of evidence that also points in the direction of increasing functional specialization comes from the analysis of the activity areas at the different sites studied. Indeed, in this sense we see that pre-Hispanic mining operations at San José del Abra underwent a transformation from minimum spatial segregation of their activity areas during the Late Formative Period (and even some superimposition), to the opposite situation during the Late Period, when primary and secondary crushing are delimited by stone wall structures, and separated from each other by a distance of about 50 meters, and the dry stream bed at the bottom of a gulley.

It is the latter evidence that indicates that an increasing specialization took place not only in terms of the search for tools well-suited to overcome physical restrictions, but also in terms of workers dedicated more specifically to each phase of the productive process, at least during the Late Period. The Late Intermediate Period, on the other hand, is an intermediate situation, somewhere between that which is seen during immediate prior and later times of the chronological sequence. We have to keep in mind that the populations which exploited copper minerals in San José del Abra were always atacamenian. This is especially significant during the Late Period, when this region came under Inka rule. But there is no direct evidence that the Incas brought with them technological changes that affected the millenary tradition of atacamenian mining. Neither is there evidence of Inca populations working at the mines. So we must conclude that the local atacamenian mining tradition underwent changes as a result of an increased specialization of mining activities, which at the same time was made possible by the social and economic transformations that the Inca State introduced in the region.

We suggest that the Incan Empire took advantage of the previous knowledge and technological experience that was still in the process of functional specialization, boosting this process by means of an efficient system for financing and administering the mining operation, both in terms of the available stock of instruments, as of feeding and organizing the workforce.

So technological changes cannot be understood in isolation from their social and economic contexts at a regional scale (Lechtman 1991; Núñez 1987; González 1995). Rather it is through the consideration of this regional context that the local transformations and technological may be coherently interpreted.

References

Adán, L. and M. Uribe 2005. El dominio Inka en las quebradas altas del Loa Superior: un acercamiento al pensamiento político andino. *Estudios Atacameños* 29, 41-66.

Ambrus, J. 1977. Geology of El Abra porphyry copper deposit, Chile. *Economic Geology* 72, 1062-1085.

Ayala, P. 2001. Las sociedades Formativas del altiplano Circumtiticaca y Meridional y su relación con el Norte Grande de Chile. *Estudios Atacameños* 21, 7-47.

Barba, A. 1967. *El arte de los metales*. Potosí, Colección de la Cultura Boliviana [1640].

Berenguer, J. 2004. *Caravanas, interacción y cambio en el Desierto de Atacama*. Santiago de Chile, Sirawi Ediciones.

Berenguer, J. and P. Dauelsberg 1989. El norte grande en la órbita de Tiwanaku (400 a 1.200 d. C.). In J. Hidalgo (ed.), *Culturas de Chile. Prehistoria: Desde sus orígenes hasta los albores de la conquista*. Editorial Andrés Bello, Chile, 129-180.

Boman, E. 1908. *Antiquites de la Region Andine et du Desert d'Atacama*, Paris, Imprimerie Nationale.

Craddock, B.R. 1990. The experimental hafting of stone mining hammers. In P. Crew and S. Crew (eds.), *Early mining in the British Isles*, .Plas Tan y Bwlch Snowdonia National Park Study Center, London.

Esperou, J.L., Roques, P. and P. Ambert 1992. L'outillage des mineurs préhistoriques de Cabrières : Les Broyeurs, Colloque en Hommage à J. Arnal, Saint-Mathieu de Tréviers, *Archéologie en Languedoc*, 15, 67-76.

Gerwe, J., Latorre, J. and L. Barrett 2003. El Abra porphyry copper deposit, Northern Chile. An update. In: XI Congreso Geológico Chileno, Concepción, Chile.

González, L. R. 1995. Recursos y organización de la producción metalúrgica prehispánica en la región centro-sur. Un caso de estudio. *Hombre y Desierto*, 9, 118-132.

González, L. 2004. *Bronces sin nombre, la metalurgia prehispánica en el Noroeste Argentino*, Ediciones Fundación Ceppa, Buenos Aires.

Hertzberg, R. 1976. *Deformation and fracture mechanics of engineering materials*, Ed. John Willey & Sons, New York.

Iribarren, J. 1962. Minas de explotación por los Incas y otros yacimientos arqueológicos en la zona de Almirante Latorre, Departamento de La Serena, La Serena, *Boletín del Museo Arqueológico de La Serena* 13, 61-72.

Iribarren, J. 1971. Una mina de explotación Incaica: El Salvador-Provincia de Atacama. *Boletín de Prehistoria*, 46-49.

Latcham, R. 1938. *Arqueología de la región atacameña*, Prensas de la Universidad de Chile, Santiago.

Lechtman, H. 1976. A metallurgical site survey of the Peruvian Andes, *Journal of Field Archaeology* 3 (1), 1-42.

Lechtman, H. 1980. The Central Andes: metallurgy without iron. In T. Wertime and J. Muhly (eds.), *The coming of the age of iron*, University Press, Yale, 267-334.

Lechtman, H. 1991. La metalurgia precolombina: tecnología

y valores. Los orfebres olvidados de América, Museo Chileno de Arte Precolombino, Santiago.

Llagostera, A. 1976. Hipótesis sobre la expansión incaica en la vertiente occidental de los Andes Meridionales. In J.M. Cassassas (ed.), *Homenaje al R. P. Gustavo Le Paige S.J.*, Universidad del Norte, Antofagasta, 203-218.

Moraga, A. 2000. Un modelo mineralógico en el yacimiento El Abra, tipo pórfido cuprífero. Actas del IX Congreso Geológico Chileno, vol. 1, 298-302.

Muñoz, I. 1989. El Período Formativo en el Norte Grande. Culturas de Chile. Prehistoria: Desde sus orígenes hasta los albores de la conquista. Editorial Andrés Bello, Chile, 107-128.

Niemeyer, H. and V. Schiappacasse 1988. Patrones de asentamiento incaico en el Norte Grande de Chile. In T. Dillehay and P. Netherly (eds.), *British Archaeological Reports, International Series*, BAR Publishing, Oxford, 141-179.

Núñez, L. 1987. Tráfico de metales en el área centro-sur andina: factos y expectativas. *Argentina, Cuadernos del Instituto Nacional de Antropología*, 12, 73-105.

Núñez, L. 1992. *Cultura y conflicto en los oasis de San Pedro de Atacama*, Editorial Universitaria, Santiago.

Núñez, L. 1994. Emergencia de complejidad y arquitectura jerarquizada en la Puna de Atacama: las evidencias del sitio Tulán-54. In M. E. Albeck (ed.), *Taller de Costa a Selva*, Instituto Interdisciplinario de Tilcara, Jujuy, 85-116.

Núñez, L. 1997. Informe sobre rescate arqueológico de materiales culturales de los sitios Ab-22 y Ab-43. *Geotécnica Consultores* (unpublished manuscript).

Núñez, L. 1999. Valoración minero-metalúrgica circumpuneña: menas y mineros para el Inka rey, San Pedro de Atacama. *Estudios Atacameños*, 18, 177-222.

Núñez, L. 2006. La orientación minero-metalúrgica de la producción atacameña y sus relaciones fronterizas. In H. Lechtman (ed.), *Esferas de interacción prehisptóricas y fronteras nacionales modernas: los Andes surcentrales*, IEP-IAR, Lima, 205-260.

Núñez, L., Cartagena, I., Carrasco, C., de Souza P. and M. Grosjean 2006. Emergencia de comunidades pastoralistas formativas en el sureste de la Puna de Atacama, San Pedro de Atacama. *Estudios Atacameños*, 32, 93-117.

Petersen, G. 1970. Minería y metalurgia en el antiguo Perú, Lima, Arqueológicas 12, Museo Nacional de Antropología y Arqueología. *Arqueológicas*, 12, 1-140.

Pickin, J. 1990. Stone Tools and Early Metal Minnning in England and Wales. In P. Crew and S. Crew (eds.), *Early mining in the British Isles*, Plas Tan y Bwlch Snowdonia National Park Study Center, London, 39-42.

Raffino, R. 1981. *Los Inkas del Kollasuyu*, Buenos Aires, Editorial Ramos Americana.

Ravines, R. 1978. Metalurgia. In: R. Ravinés (ed.), *Tecnología Andina*. Lima, Instituto de Estudios Peruanos, 476-487.

Rees C. 1999. Elaboración, distribución y consumo de cuentas de malaquita y crisocola durante el Período Formativo en la Vega de Turi y sus inmediaciones, subregión del río Salado, norte de Chile. In C. Aschero, M. Korstanje and P. Vuoto (eds), *Los tres reinos: prácticas de recolección en el cono sur de América*, Universidad Nacional de Tucumán, Argentina, 85-98.

Rees, C. and P. De Souza 2004. Procesos de producción lítica durante el Período Formativo en la subregión del Río Salado (II Región, norte de Chile). Actas del XV Congreso Nacional de Arqueología Chilena, Chungara, Número Especial, 453-465.

Salazar D. 2002. El Complejo minero San José del Abra, II Región. Una aproximación a la arqueología de la minería. Unpublished Master's Thesis, Escuela de Postgrado, Universidad de Chile, Santiago.

Salazar, D. 2002-2005. Investigaciones arqueológicas sobre la minería incaica en San José del Abra (II Región, norte de Chile), Xama, Mendoza, 15-18, 101-117.

Salazar, D 2003-2004. Arqueología de la minería: propuesta de un marco teórico, *Revista Chilena de Antropología*, Santiago,17, 125-150.

Salazar D. and H. Salinas 2005. Avances en el estudio de la organización tecnológica minera durante el Período Tardío en dos localidades atacameñas. Actas del XV Congreso Nacional de Arqueología Argentina, Cordoba.

Salazar, D. and H. Salinas 2007. Tradición y transformaciones en la organización de los sistemas de producción mineros en el norte de Chile prehispánico: San José del Abra, siglos I al XVI d.C. In P. Cruz, P. Absi and M. Van Buren (eds.), *Minas y metalurgia en los Andes centrales y del Sur. Desde tiempos prehispánicos hasta el siglo XVII.* Sucre, IRD/IFEA/University of Colorado State/ABNB.

Salazar D., Salinas, H. and P. Corrales 2005. El pirquén olvidado: reflexiones arqueológicas a partir de faenas pirquineras en el norte de Chile. *Actas de las V Jornadas de Arqueología e Historia de las Regiones Pampeana y Patagónica*, Luján, Argentina.

Salazar, D., Salinas, H., Mcrostie, V., Labarca, R. and G. Vega 2006. Cerro Turquesa: diez siglos de producción minera en el extremo Norte de Chile, Actas del XVII Congreso Nacional de Arqueología Chilena, Valdivia.

Salinas, H. and D. Salazar 2008. Cadenas operativas y sistemas de explotación minera prehispánica. In D. Jackson, D. Salazar and A. Troncoso (eds.) *Puentes hacia el pasado: reflexiones teóricas en arqueología*, Serie Monográfica de la Sociedad Chilena de Arqueología 1, Santiago, 73-92.

Salinas, H., Salazar D., Órdenes D. and J. Parra. 2006. Organización tecnológica y sistemas de explotación minera prehispánica. *Actas* del XVII Congreso Nacional de Arqueología Chilena, Valdivia.

Salinas, H. 2007. *Estudios acerca de la organización de la tecnología minera prehispánica en el Loa Superior, norte de Chile*. Memoria para optar al título de arqueólogo, Departamento de Antropología, Universidad de Chile.

Sepúlveda, M. and E. Laval 2010. Uso de minerales de cobre en la pintura rupestre de la localidad del Río Salado (II Región, Norte De Chile). Actas del XVII Congreso Nacional de Arqueología Chilena, Valdivia, 1111-1122.

Schiappacasse, V., Castro V. and H. Niemeyer 1989. Los desarrollos regionales en el Norte Grande (1000-1400

DC). In J. Hidalgo (eds.), *Culturas de Chile. Prehistoria*, Andrés Bello, Santiago, 181-220

Shimada, I. 1994. Pre-Hispanic metallurgy and mining in the Andes: recent advances and future tasks. In A. Craig & R. West (eds.), Quest of mineral wealth: aboriginal and colonial mining and metallurgy in Spanish America, Louisiana State University, Baton Rouge, *Geoscience and Man*, vol. 33, 37-73.

Soto, C. 2006. Tipología De Cuentas De Collar En La Quebrada De Tulan (Salar De Atacama): Nueva línea de evidencia para la transición Arcaico-Formativo. Actas del XVII Congreso Nacional de Arqueología Chilena, Valdivia.

Thornburn, J. 1990. Stone Mining Tools and the Field for Early Mining in Mid-Wales. In P. Crew and S. Crew (eds.), *Early mining in the British Isles*, Plas Tan y Bwlch Snowdonia National Park Study Center, London, 43-48.

Timberlake 1990. Excavations at Parys Mountain and Nantyreira. In P. Crew and S. Crew (eds.), Early mining in the British Isles, Plas Tan y Bwlch Snowdonia National Park Study Center, London, 15-21.

Ulricksen, C. 1990. Cuadrángulo Cerro Jaspe y sector occidental del cuadrángulo Estación San Pedro. Escala 1:50.000. Unpublished Manuscript.

Uribe, M. 1999-2000. La arqueología del inka en Chile, Santiago, *Revista Chilena de Antropología*, 15, 63-97.

Westfall, C. and C. González 2006. Mina Las Turquesas: un asentamiento minero lapidario preincaico en el extremo meridional del área circumpuneña, región de Atacama. Actas del XVII Congreso Nacional de Arqueología Chilena, Valdivia.

www.ingramcontent.com/pod-product-compliance
Lightning Source LLC
Chambersburg PA
CBHW061302270326
41932CB00029B/3441